International Series of Monographs on Physics

120. M. Warner, E. Terentjev: *Liquid crystal elastomers*
119. L. Jacak, P. Sitko, K. Wieczorek, A. Wójs: *Quantum Hall systems*
118. J. Wesson: *Tokamaks Third edition*
117. G. Volovik: *The universe in a helium droplet*
116. L. Pitaevskii, S. Stringari: *Bose–Einstein condensation*
115. G. Dissertori, I. G. Knowles, M. Schmelling: *Quantum chromodynamics*
114. B. DeWitt: *The global approach to quantum field theory*
113. J. Zinn-Justin: *Quantum field theory and critical phenomena, Fourth edition*
112. R. M. Mazo: *Brownian motion—fluctuations, dynamics, and applications*
111. H. Nishimori: *Statistical physics of spin glasses and information processing—an introduction*
110. N. B. Kopnin: *Theory of nonequilibrium superconductivity*
109. A. Aharoni: *Introduction to the theory of ferromagnetism, Second edition*
108. R. Dobbs: *Helium three*
107. R. Wigmans: *Calorimetry*
106. J. Kübler: *Theory of itinerant electron magnetism*
105. Y. Kuramoto, Y. Kitaoka: *Dynamics of heavy electrons*
104. D. Bardin, G. Passarino: *The Standard Model in the making*
103. G. C. Branco, L. Lavoura, J. P. Silva: *CP violation*
102. T. C. Choy: *Effective medium theory*
101. H. Araki: *Mathematical theory of quantum fields*
100. L. M. Pismen: *Vortices in nonlinear fields*
99. L. Mestel: *Stellar magnetism*
98. K. H. Bennemann: *Nonlinear optics in metals*
97. D. Salzmann: *Atomic physics in hot plamas*
96. M. Brambilla: *Kinetic theory of plasma waves*
95. M. Wakatani: *Stellarator and heliotron devices*
94. S. Chikazumi: *Physics of ferromagnetism*
91. R. A. Bertlmann: *Anomalies in quantum field theory*
90. P. K. Gosh: *Ion traps*
89. E. Simánek: *Inhomogeneous superconductors*
88. S. L. Adler: *Quaternionic quantum mechanics and quantum fields*
87. P. S. Joshi: *Global aspects in gravitation and cosmology*
86. E. R. Pike, S. Sarkar: *The quantum theory of radiation*
84. V. Z. Kresin, H. Morawitz, S. A. Wolf: *Mechanisms of conventional and high T_c superconductivity*
83. P. G. de Gennes, J. Prost: *The physics of liquid crystals*
82. B. H. Bransden, M. R. C. McDowell: *Charge exchange and the theory of ion-atom-collision*
81. J. Jensen, A. R. Mackintosh: *Rare earth magnetism*
80. R. Gastmans, T. T. Wu: *The ubiquitous photon*
79. P. Luchini, H. Motz: *Undulators and free-electron lasers*
78. P. Weinberger: *Electron scattering theory*
76. H. Aoki, H. Kamimura: *The physics of interacting electrons in disordered systems*
75. J. D. Lawson: *The physics of charged particle beams*
73. M. Doi, S. F. Edwards: *The theory of polymer dynamics*
71. E. L. Wolf: *Principles of electron tunneling spectroscopy*
70. H. K. Henisch: *Semiconductor contacts*
69. S. Chandrasekhar: *The mathematical theory of black holes*
68. G. R. Satchler: *Direct nuclear reactions*
51. C. Møller: *The theory of relativity*
46. H. E. Stanley: *Introduction to phase transitions and critical phenomena*
32. A. Abragam: *Principles of nuclear magnetism*
27. P. A. M. Dirac: *Principles of quantum mechanics*
23. R. E. Peierls: *Quantum theory of solids*

THE
INTERNATIONAL SERIES
OF
MONOGRAPHS ON PHYSICS

Quantum Hall Systems

Braid groups, composite fermions, and fractional charge

LUCJAN JACAK
PIOTR SITKO
KONRAD WIECZOREK
ARKADIUSZ WÓJS

Institute of Physics, Wrocław University of Technology, Wrocław, Poland

OXFORD
UNIVERSITY PRESS

This book has been printed digitally and produced in a standard specification in order to ensure its continuing availability

OXFORD
UNIVERSITY PRESS

Great Clarendon Street, Oxford OX2 6DP

Oxford University Press is a department of the University of Oxford. It furthers the University's objective of excellence in research, scholarship, and education by publishing worldwide in

Oxford New York

Auckland Cape Town Dar es Salaam Hong Kong Karachi Kuala Lumpur Madrid Melbourne Mexico City Nairobi New Delhi Shanghai Taipei Toronto

With offices in

Argentina Austria Brazil Chile Czech Republic France Greece Guatemala Hungary Italy Japan South Korea Poland Portugal Singapore Switzerland Thailand Turkey Ukraine Vietnam

Published in the United States by Oxford University Press Inc., New York

Reprinted 2008

ISBN 978-0-19-852870-8

ACKNOWLEDGEMENT

The authors would like to acknowledge Professor John J. Quinn as coauthor of original ideas of Chapters 8 and 9.

PREFACE

The presented book covers some important topics in the theory of physics of quantum Hall effects. The discoveries of both quantum Hall effects – of the integer one and of the fractional one were awared Nobel Prizes, in 1985 (Klaus von Klitzing) and in 1998 (Robert B. Laughlin, Daniel C. Tsui and Horst L. Störmer), respectively. In particular, Robert B. Laughlin was awarded for his theory of the fractional quantum Hall effect (introducing the Laughlin wave function and fractionally charged quasiparticles – obeying fractional statistics).

The special attention in the book is paid to the fractional quantum Hall effect and its connections with topology of planar systems. The book starts with the fundamental problems of quantum statistics in two dimensions and the corresponding braid group formalism. The braid group theory of anyons, previously known, is developed for composite fermions. The main formalism used in many-body quantum Hall theories – the Chern-Simons theory is presented. The Chern-Simons theory of anyons obeying fractional statistics and composite fermions, related to quantum Hall systems, is given in detail. Numerical studies, which play the important role in quantum Hall theories, are presented for spherical systems (Haldane sphere). The composite fermion theory is tested in numerical studies. The concept of the hierarchy of condensed states of composite fermion excitations is introduced in analogy to the Haldane hierarchy (first condensed states of composite fermion excitations have been very recently found in the experiment by Tsui-Störmer group in 2003, which again manifested the nontrivial topology character of quantum Hall systems). The hierarchies of odd-denominator states and even-denominator states are presented. The BCS paired Hall state and the related nonabelian braid statistics are also discussed. The introduction into multi-component quantum Hall systems and spin quantum Hall systems is sketched. The theory is presented at the level which should be accessible also for graduate students in physics.

Quantum Hall effects were discovered more than 20 years ago. Despite the long time passing from discoveries of the effects there are still unanswered questions and unsolved problems. Also, the experiment still gives data which are really suprising and unexpected. We think our book may help to shed some new light into the field. Especially, as it shows deep connections between quantum Hall physics and the topology of two-dimensional many-body systems. We hope the book may give the reader deeper insight of the underlying physics and we belive it can serve as a good starting point for studying quantum Hall systems still being one of the most intriguing area of the condensed matter physics.

CONTENTS

1

INTRODUCTION

The discoveries of the quantum Hall effects, both fractional (Tsui *et al.* 1982) and integral (von Klitzing *et al.* 1980), opened a new area in physics of two-dimensional electron systems. Both discoveries were awarded Nobel Prizes, for the integer quantum Hall effect—Klaus von Klitzing in 1985; for the fractional quantum Hall effect—Horst L. Störmer, Daniel C. Tsui, and Robert B. Laughlin in 1998. In the case of the integer quantum Hall effect (IQHE), conductivity becomes quantized and appears to be the integer multiplicity of e^2/h—a simple combination of fundamental constants (Laughlin 1981; Prange and Girvin 1987; Das Sarma and Pinczuk 1997). This effect does not depend on specific parameters of the material but is connected with the Landau quantization of electron dynamics in the magnetic field, which in the case of two-dimensional systems leads to the quantization of conductivity (it is important to add, however, that impurities and boundary conditions, which remove the Landau levels degeneracy, play a vital part in the quantum Hall effect (Prange and Girvin 1987)). The fractional quantum Hall effect (FQHE) has appeared to be an absolutely unexpected phenomenon, which cannot be explained simply in the light of the Landau levels structure. The essence of the FQHE is the occurrence of characteristic anomalies in the conductivity at magnetic fields corresponding to the fractionally filled lowest Landau level (fractions with an odd denominator). In the case of the two-dimensional system of electrons, a fundamental analysis of quantum mechanics of low-dimensional systems is needed. Such systems offer an exotic physics, much richer than the three-dimensional ones. This richness reflects the complexities of the topological structure of the two-dimensional space. In order to explain the unexpected phenomenon of the FQHE, the new types of particles are introduced—composite fermions and anyons—as they are realized in the planar geometry of fractional quantum Hall systems.

The experiments conducted in the mid 1990s (Du *et al.* 1993; Kang *et al.* 1993; Willett *et al.* 1993; Goldman *et al.* 1994) not only proved the usefulness of composite fermion description (Heinonen 1998), but also confirmed that composite fermions can indeed be perceived as a quantum-mechanical consequence of topological singularity of a two-dimensional space. In the case of a two-dimensional gas, for example, in Ga(Al)As hetero-structures, at the magnetic field (of order of 20 T) corresponding to the half-filling of the lowest Landau level, the magnetic field is neutralized by the internal electron structure in such a way that, finally, charged electrons in this magnetic field move like free particles in a field-free space (Halperin *et al.* 1993; Willett *et al.* 1993). Those electrons constitute a Fermi sea, of radius described by its density within the following two-dimensional relation: $p_F = \sqrt{4\pi\rho}$, which is different from the respective Fermi radius for $B = 0$ by $\sqrt{2}$ factor, which is connected with the complete spin polarization at field $B \approx 20\,\mathrm{T}$ (for which the filling of the lowest Landau level equals 1/2). This unusual Fermi liquid has some additional properties, for example, a singularity in the effective mass at the Fermi energy (Halperin *et al.* 1993; Du *et al.* 1994; Manoharan *et al.* 1994). Furthermore, the minima of the resistance tensor component R_{xx}, appearing at the FQHE hierarchy fillings (fractional fillings with odd denominators), can be explained in a quite new and natural way: they are the consequence of Shubnikov–de Haas oscillations in the magnetic field being the deviation from the field at the half-integer filling—see Fig. 1.1.

The related phenomena can be found at 5/2 and at 7/2 (also at recently observed 3/8 and 3/10 states (Pan *et al.* 2003)—the hierarchical even denominator states—see Section 8.3)—the exceptional quantum Hall even denominator states (see Fig. 1.2—here we include spins and at the filling 5/2 the lowest Landau level is occupied by electrons of both spins and the next level is half-filled). It is belived that the 5/2 (7/2) state is the state of paired composite fermions (Greiter *et al.* 1991; Moore and Read 1991). The BCS paired state of composite fermions (Greiter *et al.* 1991; Bonesteel 1999) was confirmed within numerical studies in which the Pfaffian state was tested (Pfaffian is the form of the BCS wave function in real space) (Greiter *et al.* 1991; Moore and Read 1991; Morf 1998). In the

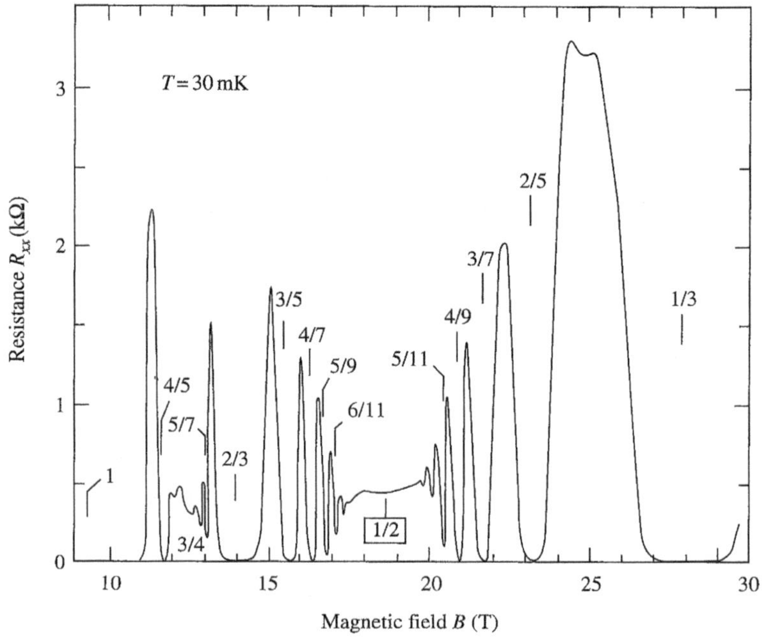

FIG. 1.1. Shubnikov–de Haas oscillations in the magnetic field $B - B_{1/2}$ (two-dimensional electron gas in Ga(Al)As), from Du *et al.* (1994).

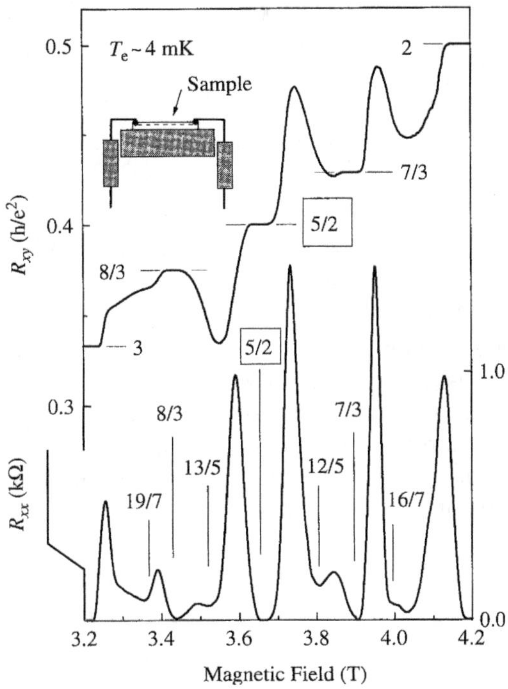

FIG. 1.2. The quantum Hall effect at the filling 5/2, from Pan *et al.* (1999*b*).

case of half-filling of higher Landau levels (e.g. at fillings $\nu = 9/2, 11/2, \ldots$) an anisotropic magnetoresistance is observed (particle density forms stripes) (Pan *et al.* 1999*a*; Rezayi and Haldane 2000).

Despite the very interesting character of the discoveries described above, it has still been unclear what happens that the electrons can neutralize the external magnetic field. This discovery also throws some new light on the very nature of the magnetic field. A question appears whether two-dimension electrons, in the absence of the external field, can also create a 'statistical' magnetic field themselves if they could neutralize an external field with its help (it would be a very strong field, tuned by electrons density, which can be easily controlled). Therefore, it seems important to analyse the mechanism of the fictitious self-induced magnetic field appearance in two-dimensional systems on the ground of recently developed theories of anyons and composite fermions.

Let us start with the following, standard quantization of states of a charged particle moving on a plane in a magnetic field, the so-called Landau quantization (Abrikosov 1972; Landau and Lifszyc 1979). Such a problem is described by the following Hamiltonian:

$$H = \frac{1}{2m}\left(-i\hbar\Delta - \frac{e}{c}\mathbf{A}\right)^2. \tag{1.1}$$

Assuming the vector potential gauge $\mathbf{A} = (0, Bx)$, the Hamiltonian (1.1) formally describes the harmonic oscillator with the centre of the oscillator shifted proportionally to p_y

$$H = -\frac{p_x^2}{2m} + \frac{1}{2m}\left(p_y - \frac{e}{c}Bx\right)^2. \tag{1.2}$$

Thus, oscillator states $\epsilon_n = \hbar\omega_c(n + 1/2)$ are obtained, where $\omega_c = eB/mc$ denotes a cyclotron frequency. All the states are equally degenerated which corresponds to the fact that the centre of the oscillator can be placed in different positions along a given sample. This degeneration equals $\Phi^{ex}/(hc/e)$ and corresponds to the number of flux quanta within the flux Φ^{ex} passing through the system surface (we consider only non-superconductive systems, where hc/e is a quantum of magnetic flux).[1] When considering the same problem in the cylindrical gauge symmetry (Landau and Lifszyc 1979), equivalent states (though numbered in different order) are obtained with energies:

$$\epsilon_{n,m} = \hbar\omega\left(n + \frac{|m| - m + 1}{2}\right). \tag{1.3}$$

The lowest state with $n = 0$ and $m = 0, 1, 2, \ldots$ is described by the following functions:

$$\phi_m = \left(\frac{1}{2\pi^2 2^m m!}\right)^{1/2}\left(\frac{z}{l}\right)^m \exp\left(-\frac{|z|^2}{4l^2}\right), \tag{1.4}$$

where $2\pi l^2 B = hc/e$, and $z = x + iy$ denotes particle coordinates in complex number notation.

Considering a many-particle state of N fermions in the lowest Landau level (i.e. with $n = 0$) and constructing the Slater determinant, we obtain

$$\Psi(z_1, \ldots, z_N) = \mathrm{const}\begin{vmatrix} 1 & z_1 & z_1^2 & \cdots & z_1^{N-1} \\ \cdots & \cdots & \cdots & \cdots & \cdots \\ 1 & z_N & z_N^2 & \cdots & z_N^{N-1} \end{vmatrix} \exp\left(-\sum_j^N \frac{|z_j|^2}{4l^2}\right). \tag{1.5}$$

The determinant in eqn (1.5) becomes the Vandemonde determinant and equals:

$$\prod_{j<k}^{N}(z_j - z_k). \tag{1.6}$$

[1]In the case of a superconductor of BCS type, $hc/2e$ is the flux quantum because the charge of the Cooper pair equals $2e$.

The famous idea of Laughlin, which introduces composite fermions, is based on replacing the above factor with the Jastrov factor (Laughlin 1983*b*):

$$\prod_{j<k}^{N}(z_j - z_k)^{2p+1}, \tag{1.7}$$

where p denotes a positive integer. Such a factor does not change the antisymmetry of the wave function. For a planar system, though, the introduction of the Jastrov factor brings a significant change: when one particle is interchanged with another, the phase of the wave function changes by $(2p+1)\pi$, and not by π. Although the factors $e^{i\pi} = e^{i(2p+1)\pi} = -1$ still denote fermions, they are topologically quite different; they suggest the existence of a source of an additional phase change. In an attempt to explain this additional phase change, the idea of Aharonov–Bohm effect can be used. On the basis of this effect, it can be assumed that $2p$ quanta of magnetic flux has been attached to each electron.[2] Thus, we obtain a composite fermion as a particle and a magnetic field localized on it. A simple analogy between the square of the Laughlin function and a classical distribution function for a two-dimensional plasma shows that a state described by this function relates to the $1/(2p+1)$ filling of the lowest Landau level[3] (Yoshioka 1985; Prange and Girvin 1987).

At first, Laughlin's idea referred only to the $1/(2p+1)$ fillings. In this case, it has proved to be a very good approximation of the system ground state, and furthermore, it has been validated in various numerical analyses (Prange and Girvin 1987). However, for the quantum Hall effect at other filling fractions with an odd denominator, a so-called 'hierarchical theory' has been developed (Yoshioka 1985). This theory consists of the following assumption: in the same way as the $1/(2p+1)$ Laughlin states of electrons appear, in a way, favoured,

[2]For two particles of charges q_1, q_2 and fluxes (i.e. localized magnetic field fluxes) Φ_1, Φ_2, their double replacement causes the phase factor $e^{i(q_1\Phi_1+q_2\Phi_2)}$ to appear in the wave function (Aharonov–Bohm effect).

[3]The Laughlin wave function gives the probability density of the system configuration:

$$|\Psi_{1/(2p+1)}(z_1,\ldots,z_N)|^2 = C\exp\left[-\frac{1}{2p+1}\left(2(2p+1)^2\sum_{j<k}\ln|z_j - z_k| + \frac{2p+1}{2}\sum_{j}^{N}\frac{|z_j|^2}{l^2}\right)\right].$$

Formally, it is the same expression as the classical Boltzmann distribution function for the two-dimensional homogeneous plasma of particles with $2p+1$ charge and temperature $T = (2p+1)/k$, where k is the Boltzmann constant. The logarithmic part is connected to the Coulomb interaction among particles in two-dimension; the other part—the exponent argument—denotes the interaction of negatively electrically charged particles with the positively charged background of a homogeneous charge of density $1/(2\pi l^2)$. For classical plasma, the equilibrium condition is connected to the density configuration for particles of $1/(2\pi l^2)1/(2p+1)$ density. That is why the configuration described with the wave function $\Psi_{1/(2p+1)}$ is connected to the same density of particles and refers to $1/(2p+1)$ filling of the lowest Landau level.

a possibility of a creation of similar condensed states of quasi-particles (excitations on $1/(2p+1)$ states) arises. Such quasi-particles obey fractional quantum statistics; they are neither bosons nor fermions, but anyons. The major difficulty in the hierarchical theory lies in the fact that for some experimentally observed fractions as many as four generations of quasi-particles need to be created (ones generated on top of others), which is not too much probable (Jain 1989, 1994). Therefore, in the search of a new model, Jain has formulated his idea referring to the composite fermions concept (Jain 1990, 1994). Jain's idea is a very simple one, actually. It emphasizes the new implementation of fermion statistics, which is so essential for the Laughlin wave function. Let us write the Laughlin wave function as follows:

$$\Psi_{1/(2p+1)} = \sum_{i<j} (z_i - z_j)^{2p} \Psi_1, \tag{1.8}$$

where $\Psi_{1/(2p+1)}$ denotes the Laughlin function referring to $1/(2p+1)$ filling in FQHE, Ψ_1 denotes a system wave function for the completely filled lowest Landau level.

Wave functions Ψ_1 and $\Psi_{1/(2p+1)}$ have a lot of features in common. First of all, they describe incompressible systems[4] (with an energy gap). According to Jain, the same can be done for higher Landau levels and be written as:

$$\Psi_{n/(2pn+1)} = \sum_{i<j} (z_i - z_j)^{2p} \Psi_n, \tag{1.9}$$

where Ψ_n is the wave function of the system with n completely filled Landau levels. The fraction on the left side of eqn (1.9) can be easily determined when we notice that multiplying by Jastrov factor means 'attaching' an even number of magnetic flux quanta $(2p(hc/e))$ to the electrons. We obtain composite fermions for which the wave function phase changes by $(2p+1)\pi$, and not by π, when the particles are interchanged. Averaging point-like magnetic fluxes (mean field approximation), we get average 'statistical' field, $B^{\mathrm{s}} = -2p(hc/e)\rho$, oriented in a direction opposite to the external magnetic field B^{ex}, thus the effective field acting on electrons equals: $B^* = B^{\mathrm{ex}} + B^{\mathrm{s}}$. Assuming that an analog of the IQHE can be observed when in the effective field n Landau levels are fully filled (Ψ_n), the FQHE at fillings $n/(2pn+1)$ (with respect to the external field) can be explained.[5] When there are two particles on a plane, charged q_1, q_2, with magnetic fluxes ϕ_1, ϕ_2 concentrated on these particles and 'attached' to them, then a change of the wave function phase for this pair, in the case of double

[4]Here, incompressibility means that it is impossible to change the area of the system (with a given number of particles) in the presence of a given external magnetic field (corresponding to a given fraction), when this area changes it leads to a change in the degeneration of Landau levels and in the case of the fully filled lowest Landau level, the particles would have to jump over the energy gap.

[5]When one assumes that the external field and the effective field are oppositely directed, the filling in the external magnetic field equals $n/(2pn-1)$.

interchange[6] of positions of the particles, is described by $q_1\phi_1 + q_2\phi_2$. For equal charges $q_1 = q_2 = q$ and fluxes $\phi_1 = \phi_2 = \phi$, the phase change is $2q\phi$. The rule describing the change of the phase of the wave function for two-dimensional system is connected by the following product *charge* $\times$ *flux*. Therefore, when a charge is specified, the whole problem is defined by the strength of the individual particle magnetic flux. Having the above in mind, it can be stated that fermions on the plane are the same as bosons supplied with an appropriately defined *flux*, so that when an exchange occurs, the phase changes by π. Composite fermions are bosons with such a flux that the phase change is equal to $(2p+1)\pi$. Anyons are defined as bosons supplied with such a flux that when they have been interchanged, the phase change equals $\theta \in (-\pi, \pi]$. However, it is important to bear in mind that anyons can be represented as fermions with an appropriately chosen flux, where the phase change equals $\phi = \theta - \pi$. Two-dimensional bosons can also be treated as fermions with an appropriately chosen flux which leads to the additional phase change of π (composite fermions, then, are fermions supplied with a flux which causes an additional phase shift of $2p\pi$; by the same analogy, composite bosons and composite anyons can be considered). In the light of the ideas above, the representations of individual charged particles on the plane can be changed, bearing in mind only that an appropriate $q\phi$ product value should be preserved. The change of a representations is only a formal procedure and it is called statistics transmutation (Wilczek 1990).

As theory allows existence on the plane of all fermion, boson, and anyon quantum systems, which quantum statistics is preferred in the physical situations. Physical interactions of particles seem to play a crucial role here (e.g. Coulomb interaction, perhaps boundary effects connected with the way a specific physical situation has been prepared), which on the basis of the minimum energy principle affects the choice of the type of particles. The physics of three-dimensional systems can serve as an analogy where a similar question can be asked about bosons and fermions. In such cases, a requirement of renormalization of relativistic field-theory models occurs (connected with the need for removing the infinite vacuum energy) and it leads to particles with half spin of fermion nature, and in consequence to particles with integral spin of boson nature. This relation is described by the Pauli theorem about the relationship between spin and statistics. But it seems that this spin–statistics relationship is rather of topological and not only relativistic nature, as the relativistic theory started with fundamental topological arguments. However, the division of the particles into bosons and fermions is related to the most fundamental theory describing the matter structure and interactions within. It is considered that bosons transmit the interactions (photons, bosons W and Z, gluons, and gravitons) among elementary particles such as fermions (leptons and quarks). Such an assumption is confirmed by the success of theoretical supersymmetrical considerations. It is proper to say here that the mechanism

[6]Double interchange means that one particle describes a loop around another, which can be easily noticed while observing one particle from the position where another rests.

of creating particles in the course of gauge symmetry spontaneous breaking leads to natural appearing of zero-mass Goldstone bosons, or mass-dressed Higgs bosons (when continuous symmetry is spontaneously broken in the presence of a far-range gauge field). Rich in various types of particles, quantum reality of two-dimensional physics allows a unique opportunity to test general ideas concerning the matter structure within experimentally accessible condensed phase objects, which can be related to unexplained field-theory or elementary particles problems, and for which typical, accelerator-based experiments border on unfeasibility.

What are particles carrying fluxes? For example, they can be considered in the light of the extension of the Hubbard model, where an appropriate exponential factor should be added to the amplitude of the jump between two neighbouring particle positions, so that when one particle describes a circle around the other, the phase changes according to the previously defined statistics, for example, the fractional one. Such methods, on networks, are convenient for numerical simulations in planar fractional-statistics systems, especially that they are well developed due to the common use of Hubbard models, for example, for the high-temperature superconductivity description (Altshuler *et al.* 1994).

The use of the Chern–Simons field (Arovas *et al.* 1985; Chen *et al.* 1989) leads to the most elegant and theoretically complete way of implementing non-standard statistics. It is a gauge field associated with a real particle system, which is well known in field-theory many-body systems considered in the context of, for example, superfluid ^{3}He (the Wess–Zumino element, the Volovik theory). The Chern–Simons term in Lagrangian looks as follows:

$$\Delta L = ej^{\nu}A_{\nu} + \frac{\mu}{2}\epsilon^{\alpha\eta\nu}A_{\alpha}\delta_{\eta}A_{\nu}, \tag{1.10}$$

where j^{ν} is the particle current, A is the Chern–Simons field. The equation of motion for the Chern–Simons field looks as follows:

$$ej^{\nu} = \frac{\mu}{2}\epsilon^{\nu\alpha\beta}\partial_{\alpha}A_{\beta}. \tag{1.11}$$

After integration, the anticipated relationship between a charge and a field flux for A is found.[7] After choosing an appropriate gauge, for example, $\partial_{\alpha}A_{\alpha} = 0$, the equations of motion (1.11) can be solved and the Chern–Simons field can be explicitly expressed in the following form:

$$A_{\alpha}(x) = \frac{e}{2\pi\mu}\sum_{i=1}^{N}\epsilon_{\alpha\beta}\frac{(x-x_i)_{\beta}}{\mid x-x_i\mid^2}. \tag{1.12}$$

[7]For $\nu = 0$, we get

$$Q = Ne = \int j^0\, ds = b\int \frac{\mu}{2}\epsilon^{0\alpha\beta}\partial_{\alpha}A_{\beta}\, ds = \mu\Phi.$$

Turning to the Hamilton notation, an appropriate contribution to the Hamiltonian connected with the Chern–Simons field can be determined (in the gauge div $\mathbf{A} = 0$)

$$H = \frac{1}{2m} \sum_{i=1}^{N} [\mathbf{p}_i - e\mathbf{A}(x_i)]^2. \tag{1.13}$$

As the eqns (1.12) and (1.13) show, the Chern–Simons field is strongly non-local and for free particles (anyons, composite fermions) we have to deal with a type of statistical interaction (a sort of generalized exchange interaction). This explains the fact that even for systems of free anyons or composite fermions, the single-particle approximation becomes a little far-fetched which happens quite often for strongly interacting systems (that is why the Laughlin function cannot be represented as the Slater determinant).

The Chern–Simons term is of a topological nature; it is connected to the Levi–Civita tensor (not the metric tensor, as it is in the case of the electromagnetic field). It breaks the symmetries connected with time reversal and mirror reflection. This topological distinction makes the effects connected with the Chern–Simons term independent from electromagnetic ones and, in the first approximation, insensitive to impurities or electromagnetic interaction influences, all the more that at the long-wavelength limit, in two-dimensional space, the Chern–Simons term dominates over the electromagnetic term in the Lagrangian because it contains only one derivative while the electromagnetic term two (in a three-dimensional equivalent of the Chern–Simons term there are two derivatives, which makes it comparable to the electromagnetic field in three dimension, and thus it becomes less important than it is in two-dimensional space).

Although the difference between the Chern–Simons gauge field and the magnetic field is stressed, proofs of composite fermions existence observed in experiments show that in some situations, a sort of phase transition takes place (or rather a choice of an appropriate ground state), so that the Chern–Simons field behaves like an internal magnetic field which in particular cases can interfere with a 'real' external magnetic field (e.g. it can neutralize it as in the case of composite fermions at the 1/2 filling of the lowest Landau level—see Fig. 1.1). It is this sense that the so-called anyon superconductivity should be understood (Wilczek 1990). An averaged field of fluxes leads to an effective internal magnetic field of an electron system on the plane. An appropriate statistics $\theta = \pi(1 - 1/f)$, where f is an integer, is connected with the realization of a mean field $B^{\mathrm{s}} = 2\pi\bar{\rho}/ef$, where $\bar{\rho}$ denotes an average particle density on the plane. At this field, exactly f first Landau levels become completely filled and an energy gap, similar to that in a superconductor, follows. Thus, a Meissner effect also appears. The external field does not penetrate the inside of the system (it is a sort of convention because the system is two-dimensional), because otherwise this field would modify the internal field (of a statistical nature) and would cause a forbidden energy increase.[8] Also, a gap-free collective excitation of phonon dispersion nature appears. In this case, critical velocity appears below the limit when the excitation creation is blocked by (energy and momentum) conservation rules. All this, according to Landau–Feynman argumentation (Landau and Lifszyc 1979), leads to a system superfluidity.

The above discussion (based on the localized fluxes averaging) shows that there are some difficulties in physical realization of anyons. For when we want to change a system density, it can only be done by adding at the same time at least f particles (for $\theta = (1 - 1/f)\pi$ statistics), because degeneration which depends on density through the average 'statistical' field can only be changed at the least by 1. And it all can only be done for all Landau levels simultaneously. In case the system properties are retained, only an integral multiplicity of the number f of particles can be added (taken away). Likewise the density cannot be changed with the change of the sample surface; the system in such state is considered incompressible. It is connected with the necessity of conserving the quantization of the magnetic flux also in the case of an effective Chern–Simons field (for anyons f particles carry one magnetic field quantum, which cannot be divided). Thus, adding one particle to the system leads to damaging the properties of the state. It may be the display of the essence of a finite size of a system of defined number of particles. It is worth noticing that this interpretation agrees with the approach based on Haldane generalization of Pauli exclusion principle for fermions (Haldane 1991; Wu 1994). In a system of a defined number of particles on a surface, the number of particles changes, their space (in the sense of allowed quantum states) allowed for eventual occupation can be described as follows:

$$\Delta d_\alpha = -\sum_\beta g_{\alpha\beta} \Delta N_\beta, \tag{1.14}$$

where α, β denote different types of particles present in the system, ΔN_β, change in the number of β-type particles, Δd_α, change in the space available for particles of α type. In the case of $g_{\alpha\beta} = 0$ we have bosons, adding some of them to a system does not change allowed space for other particles; for fermions $g_{\alpha\beta} = \delta_{\alpha\beta}$, for anyons $g_{\alpha\beta} = p/q\delta_{\alpha\beta}$ (from formula (1.14) it appears that $g_{\alpha\beta}$ should be a rational number). For anyons suspected of superconductivity, that is, for which the mean Chern–Simons field interferes with an external magnetic field, $g_{\alpha\alpha} = 1 - 1/f$. Consequently, the number of particles can only be changed by a multiplicity of f, so that the avaiable 'space' change, Δd, remains integral (the

[8] If b denotes external magnetic field orthogonal to the two-dimensional system, then depending on the mutual orientation of the external magnetic field b and internal statistic field B, the correction of the ground-state energy will have the form (for $B \pm b$) (Chen *et al.* 1989)

$$\Delta E = \frac{f^2 e^2}{4\pi m}\left[\frac{1}{f} bB - \left(1 \mp \frac{1}{f}\right) b^2\right].$$

In both cases, for sufficiently small external field b, energy change is described by $\Delta E > 0$ (increase).

connection between $g_{\alpha\beta}$ and the statistics type describes the formula: $g_{\alpha\alpha} = \theta/\pi$, (Jacak and Sitko 1993; Dasniéres de Veigy and Ouvry 1994).

All those unusual effects become a very interesting and important subject of research not only for shedding new light on the fundamental aspects of matter structure but also for bringing new possibilities of technical application (which is especially important in the face of the need of miniaturization in electronics; it can be said that size-diminishing leads to efficiency increase in such electronic equipment as e.g. the semiconductor laser). In this context, the physics of a quantum dot—a geometrically limited system of two-dimensional electrons of strictly restricted number—appears extremely important (a well-defined technology of obtaining such objects already exists). Theoretically, in a quantum dot, such above-mentioned non-trivial topological realizations of quantum mechanics can appear. Particle interaction decides whether one of them can be considered superior to the others; a rich structure of quantum dot energy levels appears, especially in an external magnetic field (Laughlin 1983*a*; Maksym and Chakraborty 1990). However, those exotic states have not yet been experimentally observed. Since in accessible spectroscopic measurements (far infrared), a wavelength is incomparably bigger than a dot size and only the dot centre of mass is involved in response; it is a one-particle effect not reflecting the many-particle statistical nature of the system. Shorter waves and stronger fields may, perhaps, allow experimental observation of, for example, a structure of composite fermions in a quantum dot (similarly to filling hierarchy for the fractional Hall effect—transport experiments are, in the case of the quantum dot, naturally quite complicated).

We would also like to mention a very interesting aspect of the anyon theory when it is applied to quantum information processing (Bennett and DiVincenzo 2000). In this rapidly developing field (Kitaev 1997) decoherence, that is, a loss of quantum information due to an uncontrolled interaction of a quantum circuit with an environment, plays the crucial role, as it is the major obstacle on the way towards a practical implementation of the scalable quantum computer (Bouwemeester *et al.* 2000). As the decoherence (due to interactions) has rather a local character, the idea of a storage of quantum information on unlocal medium (robust against local perturbations) is very attractive. Anyons—topological non-local objects—seem to be the promising candidates and the relevant scheme of exploiting of anyons in defining a qubit has recently been suggested by Kitaev (1997). That idea, rather theoretical, and far from practical realizations, is, however, qualitatively new and is probably the most interesting idea on the way to a deeper understanding of the nature of quantum information and its decoherence. There is still a growing interest in the subject (Freedman *et al.* 2001; Averin and Goldman 2002; Yang *et al.* 2002) and eventually the anyon quantum computing will help in the future development of the desired new quantum technology.

It should be added that in planar geometry, typical for three-dimensional space spin and momentum, quantization disappears (Wilczek 1990). In two-dimensional space, all rotations commute and are described by a group $\mathcal{U}(1)$ (group of complex numbers of module 1), which leads to a continuous representation connected with the spin in two-dimensional geometry. As there is a coincidence of rotation at an angle equal to π around the mass centre of a pair of particles and exchange of particles positions on the plane, the rotation group representation and the braid group representation (related to particle interchanges and thus describing the statistics) coincide. It suggests spin and statistics relationship which has nothing in common with the relativity theory. However, relativistic quantum theory of anyons or composite fermions, which is hoped to be elaborated soon, should bring new elements enhancing understanding of the above-mentioned relationship. Lack of spin quantization on the plane brings up a question about the spin of particles that are observed in two-dimensional experiments and its relationship with the three-dimensional electron spin. Unfortunately, the question of two-dimensionality of a system itself is disputable in the light of real-life physics. In few cases only the dynamics can be treated as two-dimensional (as in QHE), when the limitations of transverse motion lead to strong restrictions of a quantum state. Likewise in electrodynamics—accurate two-dimensional electrodynamics is totally different from its three-dimensional counterpart, even in the case of its projecting on two-dimensional sub-space. It leads to various conclusions; which of them will be correct, only experiments will prove.

Our considerations are organized as follows. We start (Chapters 2–4) with the topological aspects of two-dimensional quantum many-body systems. In Chapters 5–7, the Chern–Simons many-body systems are studied with respect to anyon superconductivity and the composite fermion theory. In Chapters 8 and 9 the numerical studies (on the Haldane sphere) of FQHE systems are presented, and the composite fermion theory is tested.

2

TOPOLOGICAL METHODS FOR DESCRIPTION OF QUANTUM MANY-BODY SYSTEMS

2.1 Configuration spaces of quantum many-body systems of various dimensions

The role of topology of particle space is clearly understood in classical physics where particle paths and respective configuration and phase spaces are well defined. However, when the quantization of classical systems is considered, the topological effects have to be 'translated' into quantum mechanical 'language'. The Feynman method (path integrals) serves as the convenient approach as it covers geometrical aspects of classical paths.

Let us begin with the definition of the configuration space for N particles on the given manifold[1] M (Schulman 1968; Laidlaw and De Witt 1971). If classical (i.e. distinguishable) particles are considered, then the corresponding configuration space is the N-fold Cartesian product of M manifolds M^N. Considering the quantum indistinguishability of identical particles, one finds that the points in M^N which differ only by index permutations are identical. This leads to the orbit space[2] M^N/S_N, where S_N is the permutation group of N elements. However, the problem of singular points (for which the positions of two particles are identical) in M^N/S_N arises (these are the constant points of group S_N operating on M^N).[3] Note that the incompatibility between the configuration space of distinguishable particles and the configuration space of indistinguishable particles leads to well-known thermodynamic paradoxes (the volume $d\Gamma$ of the phase space Γ corresponds to $(1/N!)(d\Gamma/(2\pi\hbar)^{3N}))$ system states).

In quantum mechanics the global properties of the configuration space depend on boundary conditions.[4] On the other hand, the problem of singular points in M^N/S_N for which the positions of two or more particles are identical should be resolved. The presence of the singular points makes it impossible to find anything else than the Bose–Einstein statistics (Leinaas and Myrheim 1977; Dowker 1985). Thus, to avoid the above difficulty, such points are usually removed from the

[1] A manifold is a space which locally exhibits Euclidean structure, for example, a circle S^1, a sphere S^2, a torus T. All Euclidean spaces are manifolds.

[2] The orbit space M/G is a set of all orbits of the group G actions on the space M. The orbit of $m \in M$ under the action of the group G is the subset of M given by $\{gm\colon\, g \in G\}$.

[3] Such points are called diagonal points.

[4] For example, for a free particle on a straight line, the periodic boundary conditions lead to the momentum quantization and at the same time change the topology of the configuration space from the topology of the straight line R^1 into the topology of a circle S^1.

configuration space.[5] We will denote the set of diagonal points by Δ. From the fact that Δ is the set of zero measure, it is obvious that it does not influence the integrals defining the expectation values. The configuration space of the system of N indistinguishable identical particles is thus defined as:

$$Q_N(M) = (M^N \setminus \Delta)/S_N. \tag{2.1}$$

The respective space for distinguishable particles is

$$F_N(M) = M^N \setminus \Delta. \tag{2.2}$$

2.1.1 *Configuration space of a many-particle system in Euclidean space*

The explicit form of the configuration space $Q_N(M)$ can be found in some simple cases (Leinaas and Myrheim 1977). Let us limit our discussion to the k-dimensional Euclidean spaces R^k. The centre of mass in such a system is localized within the space R^k. Note that it is not true, for example, for a circle, a sphere, or a torus. Moreover, the set of all possible positions of the centre of mass fills the whole space R^k. Therefore,

$$Q_N(R^k) = R^k \times \mathcal{R}(k,\, N), \tag{2.3}$$

where R^k is a space of all possible positions of the centre of mass (which coincides with the manifold R^k) and $\mathcal{R}(k,\, N)$, called a relative space, represents relative positions of particles in the centre of mass coordinate system. Relative space $\mathcal{R}(k,\, N)$ can be obtained from $R^{k(N-1)}$ by identifying orbits of action S_N and excluding diagonal points.

For the system of two particles in Euclidean space, the explicit form of $\mathcal{R}(k, 2)$ can be determined. For such a system, the relative space is equivalent to R^k with points $\mathbf{x} = \mathbf{x}_1 - \mathbf{x}_2$ and $-\mathbf{x} = \mathbf{x}_2 - \mathbf{x}_1$ identified. The diagonal set consists of one point $\mathbf{x} = 0$. Therefore,

$$\mathcal{R}(k, 2) = R_+ \times RP_{k-1}, \tag{2.4}$$

where R_+ is a positive real line which represents the length of the vector $\mathbf{x} \in R^k$ and RP_{k-1} is the real $(k-1)$-dimensional projective space for the direction determined by the vector $\pm\mathbf{x}$ (opposite directions cannot be distinguished under action of the group S_2).

Consider a system of two particles on a line ($k = 1$). The one-dimensional particle cannot interchange positions without passing through each other and the initial ordering of particles cannot be changed. Hence, the relative space $\mathcal{R}(1, 2) = R_+$. The same result can be obtained by taking into account the fact that RP_0 is a point. Therefore, the configuration space is given by

$$Q_2(R) = R \times R_+ \tag{2.5}$$

[5] After removing of diagonal points, particles obeying Bose–Einstein statistics can still be realized. In this case, however, it can be misleading to intuitively interpret bosons as particles whose positions in the space can overlap.

and it is the half-plane without the boundary line $x_1 = x_2$ (diagonal set) described by the relation $x_2 > x_1$.

In a case of a system of two particles on the Euclidean plane R^2, the projective space RP_1 is a circle S^1 and the configuration space is given by

$$Q_2(R^2) = R^2 \times \mathcal{R}(2,2) = R^2 \times (R_+ \times S^1), \tag{2.6}$$

where R^2 represents the position of the centre of mass, and the relative space $R_+ \times RP_1$ represents two degrees of freedom of the relative motion of the particles. The relative space is equivalent to the Euclidean plane R^2 with the opposite points treated as equivalent (these points are the orbits of action of the permutation group S_2 on the relative space) without origin of the coordinate system (diagonal set consisting of a single element). It is a cone of angle 60° with a singular vertex (see Fig. 2.1).

For the system of two particles in the Euclidean space R^3, the relative space is also three dimensional. The extraction of the space R_+ from the $\mathcal{R}(2,3)$ projects the space R^3 into a sphere S^2. The exclusion of diagonal points, that is, the identification of the opposite direction, is equivalent to the identification of diametrically opposite points on the sphere S^2. In this way, the projective plane RP_2 is obtained and the configuration space is given by

$$Q_2(R^3) = R^3 \times (R_+ \times RP_2). \tag{2.7}$$

If the real space is not Euclidean, a position of the centre of mass can be localized outside the space. As an example let us consider a system of two particles moving on the circle S^1. The positions of the centre of mass fill a two-dimensional closed disc D^2 (circle S^1 is the boundary of this disc). The configuration of the system can be described using two angular coordinates ($Q_2(S^1)$ is a two-dimensional space). If β_1 and β_2 denote the pair of angular coordinates of the particles on the circle, then one could introduce a centre-of-mass angle

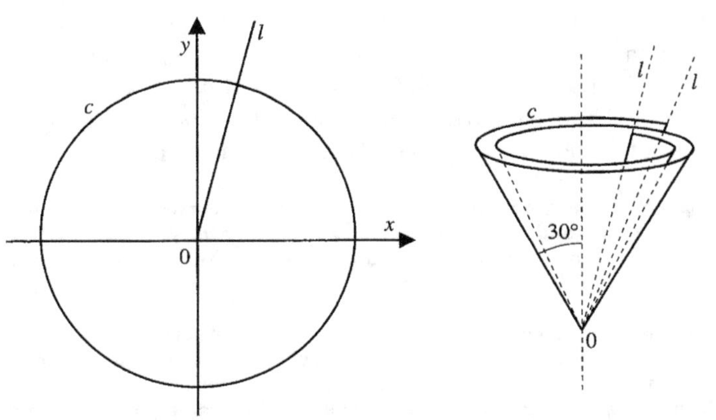

FIG. 2.1. Construction of the relative space $\mathcal{R}(2,2)$.

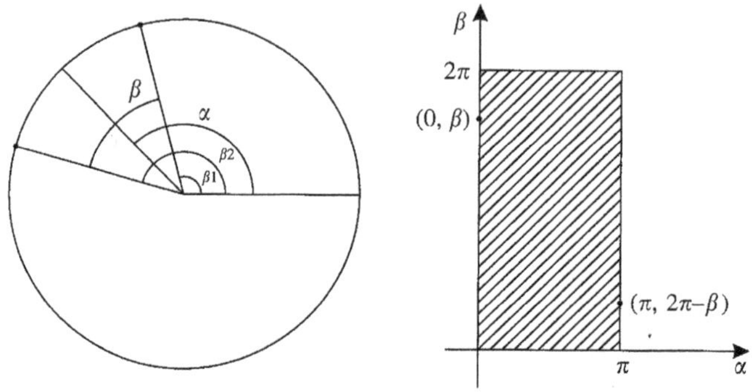

FIG. 2.2. Construction of the configuration space $Q_2(S^1)$.

$\alpha = (\beta_1 + \beta_2)/2, \alpha \in [0, \pi]$ and the relative angle $\beta = |\beta_1 - \beta_2|, \beta \in [0, 2\pi]$. Therefore, all possible coordinates fill the rectangle $J = [0, \pi] \times [0, 2\pi]$ on the plane (α, β) (see Fig. 2.2). Each point of the rectangle J is related to a different configuration of the system except for points $(0, \beta)$ and $(\pi, 2\pi - \beta)$ on the vertical boundary of J, which are equivalent. Such a boundary condition transforms the rectangle into the Möbius strip (see Fig. 2.2).

2.2 Quantization of many-body systems

The quantization language of the Feynman paths integration consists in defining the weight (measure) for various paths in the configuration space of the system. If the manifold M is a simple-connected space,[6] then the path space is a connected space (which is the consequence of the assumption that M^N is the classical configuration space).

In the case of a quantum system of indistinguishable particles, the path space is not a connected space because of the particle indistinguishability. Let us take x_1-to-x_2 paths, where x_1 and x_2 are the fixed points in the configuration space. When $x_1 = x_2$, the paths become closed and the contribution from both the paths bringing each particle into its position and the ones changing the particle positions should be taken into account. As each point in $Q_N(M)$ space relates to $N!$ points in $F_N(M)$ space, paths that are closed in $Q_N(M)$ but open in $F_N(M)$ are not homotopic[7] and the path space breaks up into separated parts, which are

[6]Simple-connected space denotes a space in which all curves which connect two different points of the space and are contained in it can be transformed one into another by means of continuous transformations. In the case of a multiply connected space, there are point-to-point curves which cannot be transformed into one another by means of continuous transformation.

[7]Two curves are homotopic when one of them can be transformed into the other by means of continuous deformations. Note that even paths which do not change the order of particles can be non-homotopic. This fact can result from the existence of topological defects in M manifold, which lead to a multiply connected space. For example, for a system of N particles in a space with a 'cut-off' hole and a path that does not change the position of particles with $1, \ldots, N-1$ indices, the path of the N-indexed particle describes a loop around the 'hole' (the

different homotopy classes.[8] The Feynman-type propagator for the transition of the system from point x at time t to point x' at time t', is defined by following equation:

$$K(xt;x't') = \sum_{\gamma} \lambda(\gamma) \int_{x(t)\in \mathcal{A}_\gamma} e^{i\int_x^{x'} Ldt} \mathcal{D}x(t), \tag{2.8}$$

where γ labels the maximal connected subspaces $\mathcal{A}_\gamma$ for path spaces of $\mathcal{A}$ system, $\lambda(\gamma)$ denotes the factor which weights the contribution of the subspace $\mathcal{A}_\gamma$ to the propagator, and L is the Lagrangian. The factor $\lambda(\gamma)$ has a quantum nature and cannot be determined by considering the classical limit (only general ideas of quantum mechanics such as linearity and unitarity can be helpful here). The homotopy theory turns out to be especially useful here. The first homotopy group π_1, called the fundamental group,[9] describes the set of topologically non-equivalent closed paths in a given space. The fact that one of the paths cannot be transformed into another one by means of a continuous deformation represents their non-equivalence. The elements of the irreducible unitary representation of the corresponding fundamental group (i.e. π_1) of the configuration space determine the contribution of individual $\mathcal{A}_\gamma$ subspaces to the propagator (Laidlaw and De Witt 1971). The choice of the particular representation will affect the propagator, because the different representations will lead to the different quantum realizations of the same classical system. The closed paths in $Q_N(M)$ represent all possible non-equivalent exchanges of particles in the given system. Thus, the first homotopy group of the configuration space of the system of identical particles refers to quantum statistics of particles under consideration.

According to the rules of paths integration, one can state that, when a closed loop is being described in the $Q_N(M)$ configuration space, then the wave function of the system must transform according to the irreducible unitary representation of the first homotopy group of this configuration space. If there are two or more irreducible unitary representations of the first homotopy group of the given configuration space, the same classical system can be quantized in many different ways (there are as many ways of quantization as there are irreducible unitary representations of the group $\pi_1(Q_N(M))$. In most cases only one-dimensional unitary representations are considered (then the wave function is a scalar). When multidimensional irreducible unitary representations of homotopy groups are considered, the wave function becomes a vector of dimension equal to the representation dimension. Such vector quantum descriptions lead to so-called parastatistics, which were once considered for three-dimensional spaces. In the case of two-dimensional manifolds, one-dimensional representations of the fundamental group (i.e. the first homotopy group) of the configuration space appear

path is closed in $F_N(M)$), this path will not be homotopic with a trivial path, for example, the one that does not change the position of any particle.

[8]The homotopy class denotes the set of all mutually homotopic curves.

[9]The fundamental group of the space Ω denotes a set of homotopy classes of the following transformations: $S^1 \rightarrow \Omega$ (Appendix A).

in such abundance that this leads to new quantum realizations without having to refer to any parastatistics.

2.3 The first homotopy group for the many-particle configuration space—braid groups

The particular configuration spaces and their homotopy groups for different manifolds which are physical spaces allowed for different many-particle systems will be analysed in the following section.

The first homotopy group of the given configuration space is often called the braid group (Birman 1974). If M denotes the manifold for the given system of N identical particles, then the group

$$\pi_1((M^N \setminus \Delta)/S_N)$$

is called the full braid group. The pure braid group, however, is the group

$$\pi_1(M^N \setminus \Delta).$$

The elements of a pure braid group denote path homotopy classes in $M^N \setminus \Delta$, for which starting and end points are identical. Classes of non-contractible closed paths for which all but one position of particles remain unchanged are the generators[10] of that group. This single particle draws a non-contractible closed path. The other particles or topological defects of the manifold M may cause the non-contractibility of this particular path. Path classes in $M^N \setminus \Delta$ whose starting and end points are connected by means of permutations of particle coordinates, that is, classes of closed paths in $(M^N \setminus \Delta)/S_N$ with fixed starting positions of the particles, are the elements of the full braid group. Thus, two types of generators of group $\pi_1((M^N \setminus \Delta)/S_N)$ are dealt with here. The first type describes the pure braid group generators, while the other one contains the classes of closed paths corresponding to exchanging two particles (without changing the positions of other particles or creating additional non-contractible loops). $\mathcal{L}$ and $\mathcal{P}$ represent those two sets of generators of $\pi_1((M^N \setminus \Delta)/S_N)$ group, respectively. $\mathcal{L}$ also denotes the set of generators for $\pi_1(M^N \setminus \Delta)$ group. Let $\Sigma_N(M)$ denote a group generated by the set of generators $\mathcal{P}$. It is easy to notice that if a manifold M is a simply-connected space, then

$$\pi_1((M^N \setminus \Delta)/S_N) = \Sigma_N(M) \tag{2.9}$$

and the group $\pi_1(M^N \setminus \Delta)$ becomes the subgroup of $\Sigma_N(M)$.[11]

If the manifold M is not a simple-connected space, then the sets $\mathcal{L}$ and $\mathcal{P}$ are disjointed and relations between their elements depend on M. In other words,

[10] The generator for the group G is an element of such subset $S \subset G$ that each element of the group G can be represented as a product of either S elements or their inverse elements and the smallest subgroup of G containing set S is the group G.

[11] If a manifold M is a simple-connected space, then generators from the set $\mathcal{P}$ generate the full braid group $\pi_1(Q_N(M))$ and all elements of the set $\mathcal{L}$.

$\pi_1(M^N \setminus \Delta)$ denotes the normal subgroup of $\pi_1((M^N \setminus \Delta)/S_N)$ for each manifold M. Moreover, the elements of the factor group $\pi_1((M^N \setminus \Delta)/S_N)/\pi_1(M^N \setminus \Delta)$ number all possible permutations of particles which can be obtained for any initial order. In most cases it is the permutation group S_N; however, it cannot be counted as a rule. Disconnected manifolds as well as a circle make exceptions. In the case of a many-particle system on a disconnected manifold M being a sum of maximal connected subspaces

$$M = \bigcup_{i=1}^{k} M_i,$$

the group

$$\tilde{S}_N = \bigoplus_{i=1}^{k} S_{N_i},$$

where N_i denotes the number of particles in the subspace N_i, will describe all possible arrangements of the system. The case of the circle will be explained later.

One should note that the space $F_N(M)$ denotes the fibration of space $Q_N(M)$ defined by epimorphism[12] h_{S_N}, connected with S_N operating on $F_N(M)$[13] (Spanier 1966; Jänich 1991). Each fibre space[14] $[F_N(M)]_x, x \in Q_N(M)$ denotes a finite space (of $N!$ elements) and a fibration is locally trivial,[15] therefore $F_N(M)$ is a covering space for $Q_N(M)$. This means that when singular points in Δ are excluded, a free (i.e. without fixed points)[16] operation of S_N on $F_N(M)$ is obtained, and homomorphism[17] from $F_N(M)$ into $Q_N(M)$ is a covering projection.[18] For such fibration, long exact sequence for a homotopy group[19] looks as

[12]Epimorphism denotes a linear mapping of one group into another.

[13]A pair consisting of a topological space B and a continuous transformation of this space into space A is called a fibration over the topological space A. In the case discussed above, it is the transformation $h_{S_N} : F_N(M) \to Q_N(M)$.

[14]An inverse image of a point x in h_{S_N} transformation is called the fibre space $[F_N(M)]_x$.

[15]Fibration h_{S_N} is locally trivial when each point $x \in Q_N(M)$ has a neighbourhood U inside which fibration is trivial, that is, $F_N(M)|U = U \times [F_N(M)]_x$, where $F_N(M)|U$ denotes the section of space $F_N(M)$ to the inverse image of U in transformation h_{S_N}.

[16]The point $p \in M$ is a fixed point of group G acting on manifold M, if there exists $g \in G$ such that $gp = p$.

[17]Homomorphism denotes a transformation $h : G \to H$ of group G into a group H such that for each $g_1, g_2 \in G$ when $g_1 \cdot g_2 = g$, then $h(g_1) \cdot h(g_2) = h(g)$.

[18]A fibration $h_{S_N} : F_N(M) \to Q_N(M)$ is covering when it is locally trivial and each fibre space $[F_N(M)]_x$ is discrete. The space $F_N(M)$ is called covering space for the space $Q_N(M)$.

[19]A sequence of homomorphisms $G_1 \xrightarrow{f_1} G_2 \xrightarrow{f_2} G_3 \xrightarrow{f_3} \dots f_{n-1} \to G_n$, where G_i are groups, is exact when a homorphism kernel f_{i+1} is the image of group G_i in homorphism f_i, $Im\ f_1 = Ker\ f_{i+1}$.

follows (Spanier 1966; Duda 1991):

$$\begin{aligned} &\longrightarrow \pi_n((F_N(M))_x, y) \longrightarrow \pi_n(F_N(M), y) \longrightarrow \pi_n(Q_N(M), x) \longrightarrow \cdots \\ &\cdots \longrightarrow \pi_1((F_N(M))_x, y) \longrightarrow \pi_1(F_N(M), y) \longrightarrow \pi_1(Q_N(M), x) \\ &\longrightarrow \pi_0((F_N(M))_x, y) \longrightarrow \pi_0(F_N(M), y), \end{aligned} \tag{2.10}$$

where M is a path-connected manifold of dimension $\dim M \geq 2$, $\pi_n(\Omega, \omega)$ is the nth homotopy group[20] of a space Ω with a base point $\omega \in \Omega$, while $x \in Q_N(M)$ and $y \in (F_N(M))_x$ are chosen arbitrarily.

Each fibre space $(F_N(M))_x$ is discrete (contains $N!$ elements), therefore, $\pi_1((F_N(M))_x, y) = \mathcal{E}$ and $\pi_0(F_N(M)_x, y) = S_N$. Moreover, the space $F_N(M)$ is path-connected which leads to $\pi_0(F_N(M)), y) = \mathcal{E}$; and for each nth homotopy group ($n \geq 1$), the introduction of the base point y can be omitted. Finally, the following sequence of homotopies is obtained (Spanier 1966; Duda 1991)

$$\mathcal{E} \longrightarrow \pi_1(F_N(M)) \xrightarrow{\alpha} \pi_1(Q_N(M)) \xrightarrow{\beta} S_N \longrightarrow \mathcal{E}, \tag{2.11}$$

where $\mathcal{E}$ is a trivial one-element group.

All generators from the set $\mathcal{L}$ must belong to the kernel of the epimorphism β, which comes from the fact that the sequence is exact. On the other hand, the generators σ_i from the set $\mathcal{P}$ are transformed into the equivalent elements of the group S_N. From Eqn (2.6) one can see that the pure braid group $\pi_1(F_N(M))$ becomes a normal subgroup of the full braid group $\pi_1(Q_N(M))$ and

$$\pi_1(Q_N(M))/\pi_1(F_N(M)) = S_N. \tag{2.12}$$

It means that in the case when the manner of interchange of particle positions is not crucial, the fundamental group of the configuration space becomes the permutation group S_N.

2.4 Braid groups for specific manifolds

2.4.1 *Full braid group for the Euclidean space R^2*

The full braid group for R^2 was described by Artin (1947) and is called the classical Artin homotopy group (it is often denoted as B_N). The elements of this group can easily be depicted by the diagram of geometrical braids)—see Fig. 2.3. One should note that the lines in the diagram can be braided in many ways; however, single strings cannot cross themselves since we removed diagonal points. If an initial arrangement does not differ from a final one, and the braid forms a closed curve in the space $F_N(R^2)$, then this braid cannot be unbraided (without cutting strings of a braid).

[20] In parallel with the case of the first homotopy group, the nth homotopy group $\pi_n(\Omega, \omega)$ is the set $[S^n, s; \Omega, \omega]$ of all homotopy classes of a transformation $h : S^n \to \Omega$ for which $\omega = h(s)$, $\omega \in \Omega$, $s \in S^n$ (Appendix A).

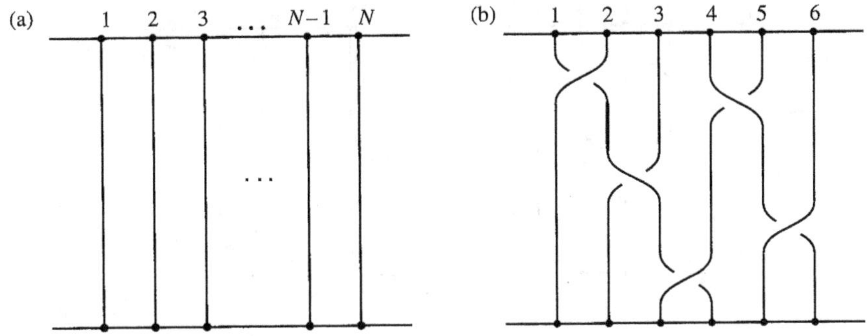

FIG. 2.3. Examples of geometrical braids: (a) trivial N-braid and (b) non-trivial 6-braid.

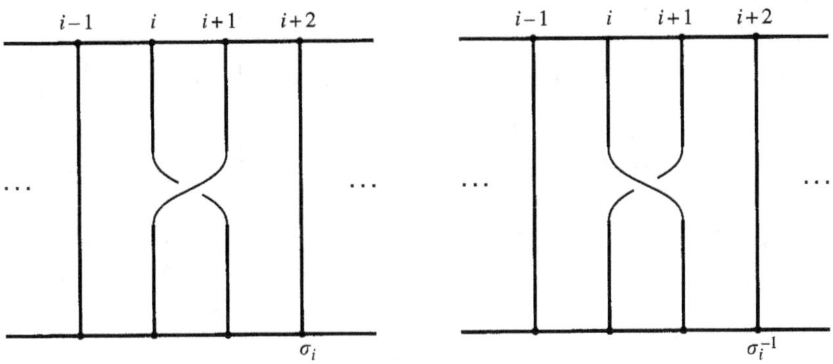

FIG. 2.4. Graphical representations of elementary operations of the exchange of particles σ_i and σ_i^{-1}.

As R^2 is a simple-connected space and $\pi_1(R^2) = \mathcal{E}$, so that the group B_N is generated by the generators from the set $\mathcal{P}$, that is, through exchanges of neighbouring particles (such braids, σ_i, are shown in Fig. 2.4) and

$$B_N = \Sigma_N(R^2). \tag{2.13}$$

The equations defining the generators σ_i are:

$$\sigma_i \cdot \sigma_{i+1} \cdot \sigma_i = \sigma_{i+1} \cdot \sigma_i \cdot \sigma_{i+1}, \tag{2.14}$$

for $1 \leq i \leq N-2$, and

$$\sigma_i \sigma_j = \sigma_j \sigma_i, \tag{2.15}$$

for $1 \leq i, j \leq N-1$, $|i-j| \geq 2$. Figure 2.5 illustrates these relations. Figure 2.6 shows the braid σ_i^2 corresponding to a double exchange of two neighbouring particles. The i and $i+1$ strings intertwine and that is why this braid is not a trivial one. It means that a double exchange of

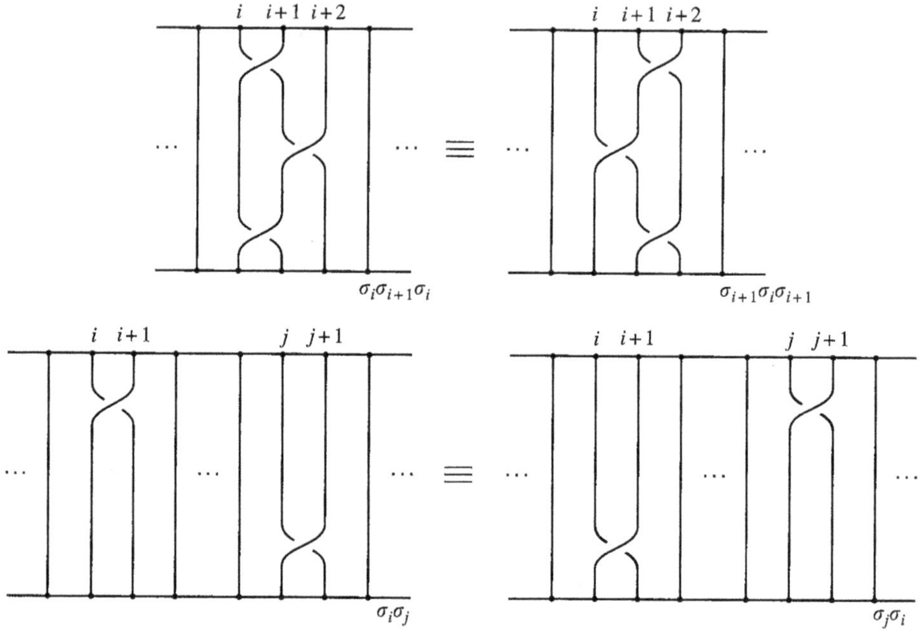

FIG. 2.5. Graphical representations of relations defining the generators σ_i of the braid group B_N.

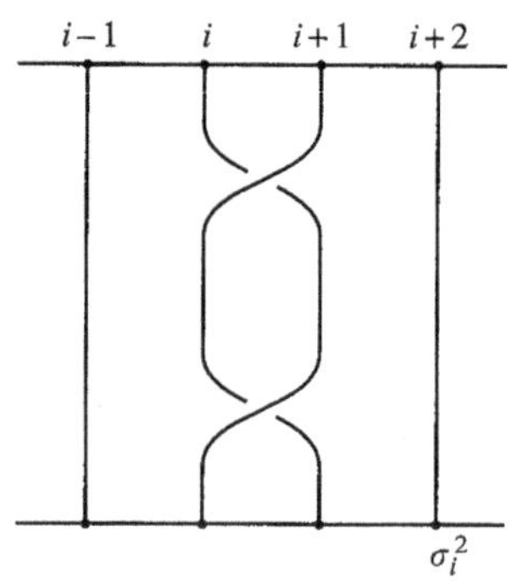

FIG. 2.6. Graphical representation of the braid σ_i^2.

particles does not lead to an identity element of the braid group B_N, that is, $\sigma_i^2 \neq e$ (in spite of the fact that σ_i^2 does not change the arrangement of particles). As one can see, in the case of the group B_N not only the starting and final arrangements of particles in the braid are important (starting and end points of a path in the space $F_N(M)$), but also the path itself.

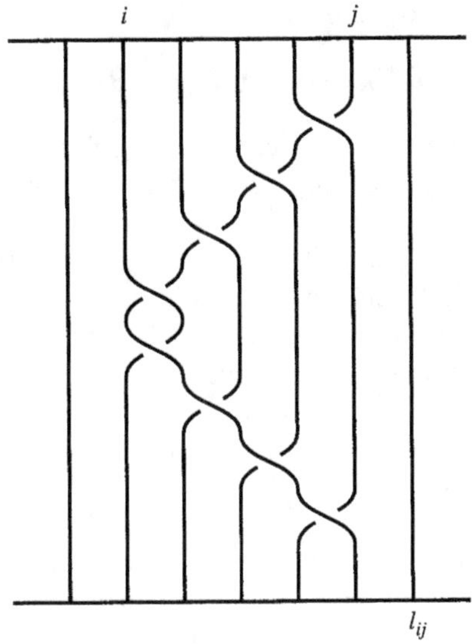

FIG. 2.7. Graphical representation of the generator l_{ij} for the pure braid group $\pi_1(F_N(R^2))$.

As $\pi_1(F_N(R^2)) \subset \Sigma_N(R^2)$, the generators from the set $\mathcal{L}$ (generating $\pi_1(F_N(R^2))$) can be thus represented by means of σ_i:

$$l_{ij} = \sigma_{j-1} \cdot \sigma_{j-2} \cdot \cdots \cdot \sigma_{i+1} \cdot \sigma_i^2 \cdot \sigma_{i+1}^{-1} \cdot \cdots \cdot \sigma_{j-2}^{-1} \cdot \sigma_{j-1}^{-1}, \tag{2.16}$$

where $1 \leq i \leq j \leq N-1$ (see Fig. 2.7). The relations which define the pure braid group $\pi_1(F_N(R^2))$ are (Hansen 1989):

$$l_{rs}^{-1} \cdot l_{ij} \cdot l_{rs} = \begin{cases} l_{ij}, & \text{for } i < r < s < j, \\ l_{ij}, & \text{for } r < s < i < j, \\ l_{rj} \cdot l_{ij} \cdot l_{rj}^{-1}, & \text{for } r < i = s < j, \\ l_{rj} \cdot l_{sj} \cdot l_{ij} \cdot l_{sj}^{-1} \cdot l_{rj}^{-1}, & \text{for } i = r < s < j, \\ l_{rj} \cdot l_{sj} \cdot l_{rj}^{-1} \cdot l_{sj}^{-1} \cdot l_{ij} \cdot l_{sj} \cdot l_{rj} \cdot l_{sj}^{-1} \cdot l_{rj}^{-1}, & \text{for } r < i < s < j. \end{cases} \tag{2.17}$$

Let us now consider the factor group $B_N/\pi_1(F_N(R^2))$. The factor group structure is defined by a homomorphism whose kernel is a normal subgroup of the group $\pi_1(F_N(R^2))$, that is, it is defined by the following equation $l_{ij} = e$, in accord with Eqn (2.16), and

$$\sigma_i^2 = e, \tag{2.18}$$

for $1 \leq i \leq N-1$. Hence, the factor group is generated by the generators σ_i which fulfil the conditions (2.14) and (2.15) together with (2.18). This group is the group of permutations of N elements, S_N,

$$B_N/\pi_1(F_N(R^2)) = S_N. \tag{2.19}$$

The group of permutations of N elements is generated by the operations of exchanges of pairs of neighbouring elements, that is, by generators parallel to σ_i.

The generators of the group S_N must fulfil conditions (2.14) and (2.15) (i.e. the braids are homotopic, see Fig. 2.5, as well as the starting and the end points of both braids are identical in the space $F_N(R^2)$). In the case of the permutation group, it is the exchange of elements which is important, not the way of realization of this exchange (that is why there is no difference between the element σ_i^2 and the element e, or, in other words, there is no difference between σ_i and σ_i^{-1}). Thus, it is clear that the generators σ_i fulfil the condition (2.18).

One can find the other presentation of the classic Artin homotopy group B_N (Coxeter and Moser 1984). The new generators are $\sigma = \sigma_1$ and $a = \sigma_1 \cdot \sigma_2 \cdot \cdots \cdot \sigma_{N-1}$. The relations which define these generators are given by

$$a^N = (a \cdot \sigma)^{N-1}, \tag{2.20}$$

$$\sigma \cdot a^{-j} \cdot \sigma \cdot a^j = a^{-j} \cdot \sigma \cdot a^j \cdot \sigma, \tag{2.21}$$

where $2 \leq j \leq N/2$. The generators σ_i can be written in the form

$$\sigma_i = a^{i-1} \cdot \sigma \cdot a^{-(i-1)}. \tag{2.22}$$

In the previous presentation of the group B_N (2.14 and 2.15) the generators were related to the operations of exchanges of two neighbouring particles (generators σ_i). However, in this new presentation, two particles exchange only when they have indices 1 and 2 (the generator σ). The powers of the generator a exchange the particles into positions 1 and 2.

2.4.2 *Full braid group for the sphere S^2*

Let S^2 be a sphere in the three-dimensional Euclidean space. Thus, $\pi_1(S^2) = \mathcal{E}$ and

$$\pi_1(Q_N(S^2)) = \Sigma_N(S^2). \tag{2.23}$$

In other words, $\pi_1(Q_N(S^2))$ is generated by operations of an exchange of neighbouring particles. A sphere is locally isomorphic with R^2, therefore, the Eqns (2.14) and (2.15) are also true for a sphere. However, the global properties of a sphere differ from the ones of R^2. It results from the fact that a closed curve on a sphere can be interpreted in two different ways. For example, when a selected particle describes a loop around all other particles, such a loop can be treated as a zero-loop. This leads to another relation for the generators of the braid group, that is,

$$\sigma_1 \cdot \sigma_2 \cdot \cdots \cdot \sigma_{n-2} \cdot \sigma_{n-1}^2 \cdot \sigma_{n-2} \cdot \cdots \cdot \sigma_2 \cdot \sigma_1 = e. \tag{2.24}$$

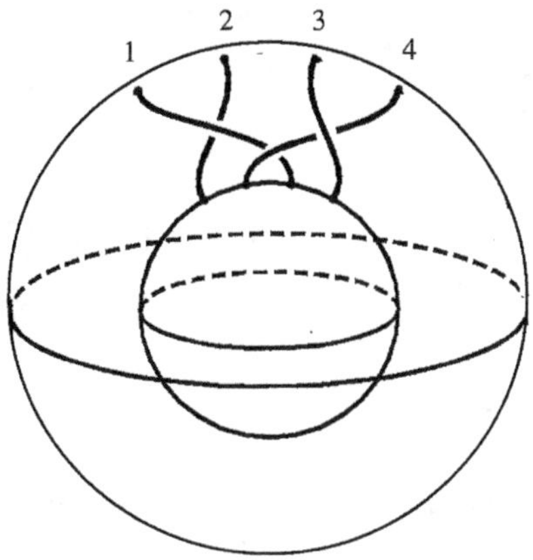

FIG. 2.8. An example of a geometrical braid for the configuration space $Q_4(S^2)$.

Figure 2.8 ilustrates this relation. Note that the relation defining generators l_{ij} for the sphere are the same as for R^2. Therefore,

$$\pi_1(Q_N(S^2))/\pi_1(F_N(S^2)) = S_N. \tag{2.25}$$

2.4.3 *Full braid group for the torus T*

A torus is a manifold which is very important to many physical interpretations as, topologically, a torus is represented in R^2 by a rectangular plaquette with periodic boundary conditions.

As $\pi_1(T) \neq \mathcal{E}$, the braid group for a torus is more complicated than the one for the sphere or the plane (Birman 1969; Einarsson 1990). Unlike a plane or a sphere, a torus is not a simple-connected manifold, therefore,

$$\pi_1(T) = \pi_1(S^1 \times S^1) = (\pi_1 S^1) \oplus \pi_1(S^1) = \mathcal{Z} \oplus \mathcal{Z} \neq \mathcal{E}, \tag{2.26}$$

where $\times$ denotes the Cartesian product, $\oplus$ denotes the simple sum, $\mathcal{Z}$ is the group of integers. Thus, $\pi_1(Q_N(T)) \neq \Sigma_N(T)$ and the set of generators $\pi_1(Q_N(T))$ divides into a set $\mathcal{L}$ generating a pure braid group $\pi_1(F_N(T))$, and a set $\mathcal{P}$ generating the group $\Sigma_N(T)$. Not all generators from the set $\mathcal{L}$ can be represented by means of a position exchange of neighbouring particles (even for a single particle there are non-contractible paths). Figure 2.9 illustrates how the generators τ_i and $\varrho_i (i = 1, 2, \cdots, N)$ operate for a pure braid group for a torus $\pi_1(F_N(T))$. The generators τ_i and ϱ_i move a particle along one of non-contractible loops on the torus while leaving all other particles in their positions. It is convenient to introduce the auxiliary generators A_{ij} and

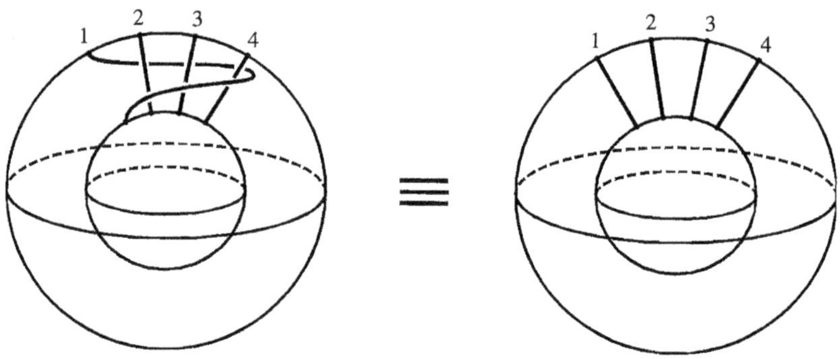

FIG. 2.9. The graphical representation of the relations defining the generators σ_i of the braid group for the sphere S^2.

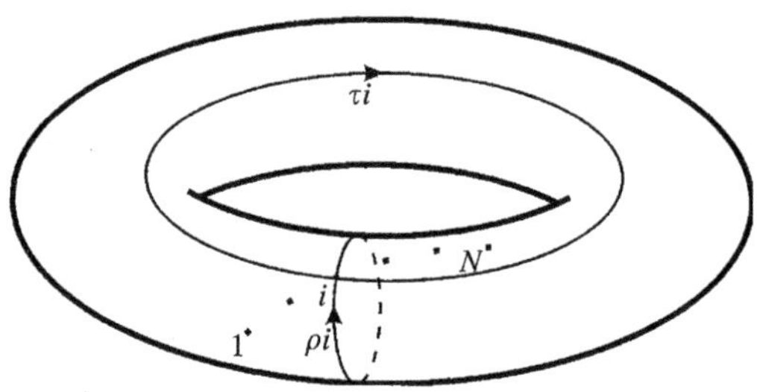

FIG. 2.10. Particle paths corresponding to the generators τ_i and ϱ_i of the pure braid group for the torus.

C_{ij}, which move the particle k around the particle l along the path shown in Fig. 2.10.

These new generators can be defined with the help of generators τ_i and ϱ_j via the following relations:

$$A_{ij} = \tau_j^{-1} \cdot \varrho_i \cdot \tau_j \cdot \varrho_i^{-1}, \tag{2.27}$$

$$C_{ij} = \varrho_j^{-1} \cdot \tau_i \cdot \varrho_j \cdot \tau_i^{-1}, \tag{2.28}$$

where $1 \leq i \leq j \leq N$.

The set of conditions which defines the relationship between the generators of the group $\pi_1(Q_N(T))$ can be divided into three subsets: the equations defining τ_i and ϱ_j, the equations defining σ_i, and the equations defining the relationship between generators from the sets $\mathcal{P}$ and $\mathcal{L}$.

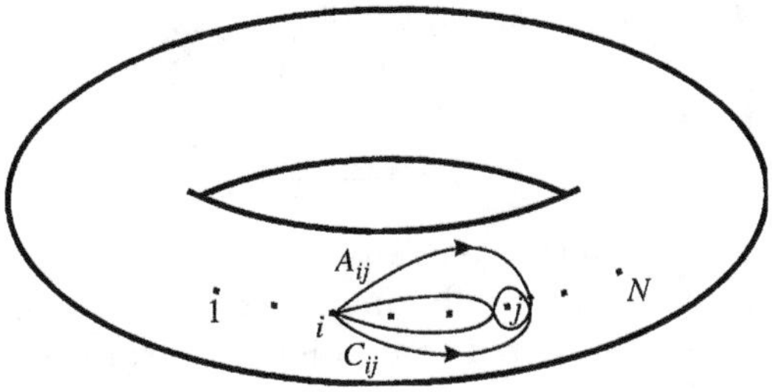

FIG. 2.11. Particle paths corresponding to the auxiliary generators A_{ij} and C_{ij}.

The following equations define the generators $\mathcal{L}$:

$$\tau_k \cdot A_{lm} = A_{lm} \cdot \tau_k, \quad \varrho_k \cdot A_{lm} = A_{lm} \cdot \varrho_k, \tag{2.29}$$

$$\tau_i \cdot \tau_j = \tau_j \cdot \tau_i, \quad \varrho_i \cdot \varrho_j = \varrho_j \cdot \varrho_i, \tag{2.30}$$

$$C_{ij} = (\tau_i \cdot \tau_j) \cdot A_{ij}^{-1} \cdot (\tau_j^{-1} \cdot \tau_i^{-1}), \tag{2.31}$$

$$A_{ij} = (\varrho_i \cdot \varrho_j) \cdot C_{ij}^{-1} \cdot (\varrho_j^{-1} \cdot \varrho_i^{-1}), \tag{2.32}$$

$$C_{ij} = (A_{j-1,j}^{-1} \cdot A_{j-2,j}^{-1} \cdot \cdots \cdot A_{i+1,j}^{-1}) \cdot A_{ij}^{-1} \cdot (A_{i+1,j} \cdot A_{i+2,j} \cdot \cdots \cdot A_{j-1,j}), \tag{2.33}$$

$$\tau_1 \cdot \varrho_1 \cdot \tau_1^{-1} \cdot \varrho_1^{-1} = A_{1,N} \cdot A_{1,N-1} \cdot \cdots \cdot A_{1,3} \cdot A_{1,2}, \tag{2.34}$$

where $1 \leq k \leq l \leq m \leq N$ and $1 \leq i \leq j \leq N$. The equations defining the generators σ_i are the same as those defining generators of the full braid group for the Euclidean plane, (2.14) and (2.15). The third set of equations, defining $\pi_1(Q_N(T))$, describes the mutual relationships between the generators from the sets $\mathcal{L}$ and $\mathcal{P}$:

$$\tau_{i+1} = \sigma_i^{-1} \cdot \tau_i \cdot \sigma_i^{-1}, \quad \varrho_{i+1} = \sigma_i \cdot \varrho_i \cdot \sigma_i, \tag{2.35}$$

$$\tau_1 \cdot \sigma_j = \sigma_j \cdot \tau_1, \quad \varrho_1 \cdot \sigma_j = \sigma_j \cdot \varrho_1, \tag{2.36}$$

$$\sigma_i^2 = A_{i,i+1}, \tag{2.37}$$

where $1 \leq i \leq N-1$ and $2 \leq j \leq N-1$.

The above examples of the homotopy groups $\pi_1(Q_N(M))$ for the configuration spaces of systems of particles on an Euclidean plane, a sphere, and a torus, show that the braid group fully describes global properties of these spaces (Mermin 1979). Although a sphere and a torus are locally isomorphic with a plane, their global properties are apparently different. In the case of a sphere, which is a closed space, there is the additional condition (2.24) which defines generators $\{\sigma_i\}$. In the case of a torus, the fact that the torus is a multiply connected space leads to two types of additional generators: τ_i and ρ_i which refer to two different fundamental non-contractible loops typical for the torus geometry.

2.4.4 *The full braid group for the three-dimensional Euclidean space R^3*

As $\pi_1(R^3) = \mathcal{E}$,

$$\pi_1(Q_N(R^3)) = \Sigma_N(R^3) \tag{2.38}$$

and this group is generated by the generators $\{\sigma_i\}$ corresponding to the exchange of i and $i+1$ particles (particle numbering is introduced). In the space $F_N(R^3)$ each closed curve can be contracted to a point (in a three-dimensional space, a point cannot be 'encircled by a loop'), thus

$$\pi_1(F_N(R^3)) = \mathcal{E}. \tag{2.39}$$

According to Eqn (2.12),

$$\pi_1(Q_N(R^3)) = S_N. \tag{2.40}$$

The Eqns (2.14) and (2.15), in addition to (2.18)[21] define the generators. The above relationships lead to essential differences between braid groups for two- and three-dimensional spaces. In the case of three dimensions, the strings shown in Fig. 2.6 do not intertwine. Moreover, each braid in a three-dimensional space can be unbraided. Therefore, when a particle transposition is considered, only the starting and final arrangements of particles are important, not the individual paths. In the case of R^3, the following equation is true:

$$\pi_1(Q_N(R^3))/\pi_1(F_N(R^3)) = S_N. \tag{2.41}$$

One can easily note that in the case of a simple-connected manifold M of dimension $\dim M > 2$, the following equation is always true[22]:

$$\pi_1(Q_N(M)) = \Sigma_N(M) = S_N. \tag{2.42}$$

When $\pi_1(M) \neq \mathcal{E}$ but $\dim M \geq 3$, then $\Sigma_N(M) = S_N$, even if $\pi_1(Q_N(M)) \neq S_N$. One can see from the above considerations that for dimensions higher or equal to three, the topology does not affect $\Sigma_N(M)$. This means that $\pi_1(Q_N(M))$ changes only when $\pi_1(F_N(M))$ does simultaneously.

2.4.5 *Braid groups for the line R^1 and the circle S^1*

The structure of homotopy groups for a configuration space for a straight line (and other one-dimensional manifolds) is different than for multidimensional spaces. This comes from the fact that there is a natural ordering of particles in this space.

In the case of R^1, the space $F_N(R^1)$ divides into equivalent simple-connected subspaces C_i which are numbered by the elements of the permutation group

[21] A double exchange of particles produces a loop when one particle moves around the other. In a three-dimensional space, one can imagine that this loop can be contracted over or below the described point.

[22] Loops described around points are always contractible in manifolds of a dimension $\dim M \geq 3$.

S_N. It means that the beginning and the end of the path belong to the same subspace C_i (it corresponds to the fact that the particle arrangement on the line is preserved). Therefore,

$$\pi_1(Q_N(R^1)) = \pi_1(F_N(R^1)) = \mathcal{E}. \tag{2.43}$$

For a circle[23] S^1 the situation becomes more complicated, because the particles can be cyclically permutated. The group $\pi_1(Q_N(S^1))$ is generated by the single generator δ which moves the first particle to the final position around the circle. The generator is of an infinite order because δ^n corresponds to a non-contractible loop, so that

$$\pi_1(Q_N(S^1)) = \mathcal{Z}. \tag{2.44}$$

The group $\pi_1(F_N(S^1))$ is generated by the generator $\nu = \delta^N$, and then $\pi_1(F_N(S^1)) = \mathcal{Z}$. Although both groups, $\pi_1(Q_N(S^1))$ and $\pi_1(F_N(S^1))$ correspond to the same abstract infinite group, the relationship between their generators leads to

$$\pi_1(Q_N(S^1))/\pi_1(F_N(S^1)) = \mathcal{Z}_N, \tag{2.45}$$

where $\mathcal{Z}_N$ is a cyclic group of order N. Similarly as two- or three-dimensional cases, this group numbers all possible arrangements resulting from the initial one.

The group Σ_N does not exist for the case of a particle system on a line and a circle because it is impossible to exchange positions of two particles without describing a non-contractible loop. Hence, the braid group formalism is not adequate for one-dimensional systems.

[23] A circle is topologically equivalent to a segment with periodic boundary conditions.

3

QUANTIZATION OF MANY-PARTICLE SYSTEMS AND QUANTUM STATISTICS IN LOWER DIMENSIONS

3.1 Topological limitations of quantum-mechanical description of many-particle systems

The quantum-mechanical description of a system with a configuration space $Q_N(M)$ consists in a construction of a time-independent state vector as a function transforming $Q_N(M)$ into a complex numbers $\mathbf{C}$ (Imbo and Sudarshan 1988; Sudarshan *et al.* 1988). This function can generally be multiple valued (e.g. it is double-valued for fermions), which results from a change of the phase factor while describing a loop in the space $Q_N(M)$. In the process of making a closed loop in $Q_N(M)$, the wave function must transform according to an irreducible unitary one-dimensional representation of the first homotopy group $\pi_1(Q_N(M))$ or to a multidimensional representation (which leads to multidimensional wave function). Therefore, in order to describe a quantization of the many-particle system, one has to choose a specific representation of the braid group describing the transformation of a state vector along the loops.

The closed paths in the space $Q_N(M)$ describe an exchange of particle positions, and therefore they are connected with the statistics of indistinguishable particles. However, the quantum statistics of particles do not always correspond to the irreducible unitary representation of the fundamental group $\pi_1(Q_N(M))$ which may be the case for a multiply connected M manifold. Even a one-particle system on a multiply connected manifold can have a complex braid group with many non-trivial representations. In such a case, a full braid group is generated by the generators $\mathcal{L}$ for the pure braid group and $\Sigma_1(M) = \mathcal{E}$. However, in such a case, it is pointless to talk about statistics. On the other hand, there are many non-equivalent realizations of quantum mechanics for such a system, but its quantum statistics should be connected not with the group $\pi_1(Q_N(M))$ but with its subgroup $\Sigma_N(M)$. The group $\Sigma_N(M)$ cannot be considered independently of the full braid group because not all of its representations will be describing distinct quantum statistics. It is possible to obtain an accurate description of quantum statistics[1] only while considering $\mathcal{W} \downarrow \Sigma_N(M)$, that

[1] In the case of multidimensional wave functions, an additional problem appears with the statistical equivalence of representations $\mathcal{W} \downarrow \Sigma_N(M)$ of different dimensions (Imbo and Sudarshan 1988; Sudarshan *et al.* 1988).

is, the limitation of the representation $\mathcal{W}$ of the group $\pi_1(Q_N(M))$ to the group $\Sigma_N(M)$.[2]

3.2 Quantum statistics and irreducible unitary representations of braid groups for selected manifolds

In this chapter, only the one-dimensional representations of the full braid group $\pi_1(Q_N(M))$ and its subgroup $\Sigma_N(M)$ will be considered. First, before determining the representation, the fundamental group should be abelized, that is, it should be divided by its commutant[3]—because one-dimensional representations are Abelian. The Abelian generators σ_i^{ab} will become the images of generators σ_i from the set $\mathcal{P}$ in the homomorphism abelizing group $\Sigma_N(M)$. The eqn (2.14) which defines the braid group generators for all two- and three-dimensional cases leads to the following equation $\sigma_i^{ab} = \sigma_j^{ab} = \sigma^{ab}$ for $1 \leq i, j \leq N-1$. That is why all one-dimensional representations of the group Σ_N will be generated by one generator; therefore the groups Σ_N will be the cyclic groups. When M is a simple-connected space, then $\pi_1(Q_N(M)) = \Sigma_N(M)$ and $\mathcal{W} \downarrow \Sigma_N(M) = \mathcal{W}(\Sigma_N(M))$. In this case the one-dimensional representations of $\Sigma_N(M)$, and simultaneously the scalar quantum statistics, will be numbered by the elements of the homomorphism group $\mathrm{Hom}([\pi_1(Q_N(M))]_{ab}, \mathcal{U}(1))$, where $\mathcal{U}(1)$ is the one-dimensional unitary group.

3.2.1 *Scalar quantum statistics of particles on the plane R^2*

For $N \geq 3$ ($B_2 = \mathcal{Z}$) and $\pi_1(F_N(R^2)) \neq \pi_1(R^2) = \mathcal{E}$, $B_N = \Sigma_N(R^2)$ is an infinite non-Abelian group on the plane. Moreover, for $N \geq 3$, the abelized group B_N is defined by:[4] $[B_N]_{ab} = \mathcal{Z}$ and $\mathrm{Hom}(\mathcal{Z}, \mathcal{U}(1)) = \mathcal{U}(1)$. The same result can be obtained when representation of generators σ_i is directly defined. These generators can be represented by any complex number denoted as $e^{i\theta}$, $\theta \in (-\pi, \pi]$, as no additional conditions defining generators σ_i exist. Each value of the parameter θ will refer to a different representation and can be treated as a number of a different quantum statistics, where $\theta = 0$ corresponds to bosons, $\theta = \pi$ to fermions, and θ different from 0 and π to particles obeying fractional statistics (anyons).

[2]Subgroup $\mathcal{W}(\pi_1(Q_N(M)))$, which is the image of the group $\Sigma_N(M)$ in the homomorphism $\pi_1(Q_N(M)) \to \mathcal{W}(\pi_1(Q_N(M)))$, is called the limitation of the representation of the group $\pi_1(Q_N(M))$ to the group $\Sigma_N(M)$ and is denoted by $\mathcal{W} \downarrow \Sigma_N(M)$.

[3]The commutant of the group G is the subgroup of the group G generated by all its so-called commutators:

$$[G, G] = \{a \cdot b \cdot a^{-1} \cdot b^{-1} : a, b \in G\}.$$

A commutant is the kernel of homomorphism that abelizes the group G, and if the group G is Abelian, then $[G, G] = \mathcal{E}$.

[4]After abelizing classical homotopy group B_N, eqn (2.14) leads to $\sigma_i^{ab} = \sigma$ for each $i = 1, 2, ..., N-1$. This means that $[B_N]_{ab}$ will be generated by one generator. As there are no any additional conditions, this generator will be of infinite order, that is, $[B_N] = \mathcal{Z}$.

It is quite unusual that in two-dimensional systems the scalar fractional quantum statistics may appear, besides Bose–Einstein and Fermi–Dirac statistics. As, in the case of the Euclidean plane, it is pointless to limit the value of the parameter θ and it can obtain any value from the interval $(-\pi, \pi]$. This means that besides the single-valued (bosons) and double-valued (fermions) wave functions, particles of other multi-valued functions may appear. Such functions are multiplied by factor $e^{i\theta}$ when two neighbouring particles are transposed. These particles are called anyons (the name was introduced by Wilczek (1982*a*)).

3.2.2 *Scalar statistics of particles on the sphere S^2*

For the sphere S^2 one has $\pi_1(Q_N(S^2)) = \Sigma_N(S^2)$. The additional condition (2.24) defining generators σ_i influences abelization[5] of group $\Sigma_N(S^2)$ and limits allowed values of the parameter θ:

$$\mathrm{Hom}(\Sigma_N(S^2), \mathcal{U}(1)) = \mathrm{Hom}([\Sigma_N(S^2)]_{ab}, \mathcal{U}(1)) = \mathrm{Hom}(\mathcal{Z}_{2N-2}, \mathcal{U}(1)), \qquad (3.1)$$

where $\mathcal{Z}_{2N-2}$ is a cyclic group with a generator of the order of $2N-2$. On the other hand, the condition $[\mathcal{W}(\sigma_i)]^{2(N-1)} = 1$ for $1 \le i \le N-1$ appears when the condition (2.24) for generators $\{\sigma_i\}$ is added. That is why the condition defining values of θ is $\theta = k\pi/(N-1)$, where $k = 1, 2, \ldots, 2N$–3. In this case, one has Fermi–Dirac statistics for $k = N-1$ and Bose–Einstein statistics for $k = 0$.

3.2.3 *Scalar statistics of particles on the torus T*

The torus T is the multiply connected manifold and the group $\pi_1(Q_N(T))$ is generated by generators of $\mathcal{P}$ and $\mathcal{L}$, and $\pi_1(Q_N(T)) \neq \Sigma_N(T)$. Therefore, the limitation $\mathcal{W} \downarrow \Sigma_N$ instead of $\mathcal{W}$ should be considered. If the set of generators $\mathcal{L}$ is abelized, then the conditions (2.27) and (2.37) lead to the condition $[\mathcal{W}(\sigma_i)]^2 = 1$ resulting in $\theta = 0$ or $\theta = \pi$. Only bosons and fermions can be realized on torus in this case. The abelization of the group $\pi_1(Q_N(T))$ leads to the following equation[6] (Imbo *et al.* 1990)

$$[\pi_1(Q_N(T))]_{ab} = H_1(T) \oplus \mathcal{Z}_2, \qquad (3.2)$$

[5]Condition (2.24), in the process of abelization, leads the new condition $\sigma^{2(N-1)} = e$ in addition to the condition $\sigma_i^{ab} = \sigma$; therefore $[\Sigma_N(S^2)]_{ab}$ becomes the finite cyclic group $\mathcal{Z}_{2N-2}$.

[6]When the braid group for the torus is abelized, the conditions (2.27) and (2.28) lead to the equation $A_{kl} = C_{kl} = e$ for $1 \le k \le l \le N$. Therefore, the conditions (2.29)–(2.35) become trivial (as well as the condition (2.36)). From eqn (2.35) we get results $\tau_k^{ab} = \tau$ and $\varrho_k^{ab} = \varrho$ for $k = 1, 2, ..., N$, and from eqn (2.14), $\sigma_i^{ab} = \sigma$ for $i = 1, 2, ..., N-1$. Thus, there is only one condition defining the set of generators $\{\tau, \varrho, \sigma\}$ for the group $[\pi(Q_N(T))]_{ab}$ and it has: $\sigma^2 = e$. The generators τ i ϱ are of infinite order and each of them generates the group $\mathcal{Z}$, and σ generates the group $\mathcal{Z}_2$. It means that $[\pi_1(Q_N(T))]_{ab} = \mathcal{Z} \oplus \mathcal{Z} \oplus \mathcal{Z}_2$, and, therefore $[\pi_1(F_N(T))]_{ab} = \mathcal{Z} \oplus \mathcal{Z}$ $[\Sigma_N(T)]_{ab} = \mathcal{Z}_2$.

where $H_1(Q_N(T))$ is the first homology group[7] of the torus, and $\mathcal{Z}_2$ is the cyclic two-element group. Moreover, the image of the group $\Sigma_N(T)$ in abelizing homomorphism is group $\mathcal{Z}_2$. It means that the representations $\mathcal{Z}_2$ describe quantum statistics of particles on the torus. This group has two representations determined by $\theta = 0$ and $\theta = \pi$. This result is true for all other closed two-dimensional manifolds (with the exception of the sphere) (Imbo *et al.* 1990).

3.2.4 *Scalar quantum statistics in three-dimensional Euclidean space* R^3

For three-dimensional Euclidean space R^3, $\pi_1(Q_N(R^3)) = \Sigma_N(R^3) = S_N$. The appearance of the additional condition (2.18),[8] for $\sigma_i^2 = e$, defining generators S_N leads to the representations equivalent to those for a torus: $\theta = 0$ and $\theta = \pi$. Therefore, in the space R^3, only boson and fermion statistics can be realized, which proves that braid group formalism leads to the experimentally observed results. For higher dimension of space as well as for additional topological string-like defects in R^3, we deal with appropriately modified pure braid group; however, different resulting non-trivial realizations of quantum mechanics will not be related to different quantum statistics (the group $\Sigma_N(M)$ remains unchanged). Therefore, in all those cases, the above-mentioned result will be true. Figure 2.6 illustrates the fact that in the plane not only boson and ferminon quantum statistics are possible. Although the element σ_i^2 does not change the ordering of particles, σ_i^2 is not the unit element of the braid group B_N (the braid strings intertwine). Along with condition $\sigma_i^2 \neq e$, the element σ_i^2 may be transformed into an irreducible unitary representation of the group $\Sigma_N(M)$ into the numbers different than one (unlike in three-dimensional Euclidean space, for which the pure braid group is denoted by $\pi_1(F_N(M)) = \mathcal{E})$). In the case of two-dimensional systems, in contrast to the three-dimensional case, $F_N(M)$ is not the universal covering space of the space $Q_N(M)$. Moreover, in a two-dimensional space, neighbouring particles can exchange positions in two ways: σ_i and σ_i^{-1} (see Fig. 2.4), while in three-dimension there is only one way for such an exchange because $\sigma_i = \sigma_i^{-1}$. The existence of non-equivalent transpositions of neighbouring particles leads to the breaking of both time reversal and spatial reflection symmetries. Time reversal changes the sense of a path, that is, in this case: σ_i^{-1} is the image of σ_i in such a transformation (see Fig. 2.4). The same result will be obtained when the generators σ_i are transformed by means of the spatial reflection symmetry (parity). As $\mathcal{W}(\sigma_i) = e^{i\theta}$ and $\mathcal{W}(\sigma_i^{-1}) = e^{-i\theta}$, both transformations lead to the exchange of θ

[7] The first abelized homotopy group of path-connected space Ω is called the first homology group $H_1(\Omega)$ of this space (Spanier 1966; Duda 1991). The first homology group, similar to the first homotopy group, is related to the transformation of a loop into the space and 'counts' the loop coils around the 'hole' in the space. Nevertheless, it is an Abelian group although it is not sensitive to the order of coils but to the total number of coils of individual 'holes'. For example, for the plane with cut out n 'holes', the first homology group is a direct sum of n groups $\mathcal{Z}$.

[8] This condition and condition (2.14) lead to $\sigma_i^{ab} = \sigma_j^{ab} = \sigma$ and condition $\sigma^2 = e$, thus $[S_N]_{ab} = \mathcal{Z}_2$.

with $-\theta$. Transformations of both time reversal and spatial reflection symmetry change the sign of the parameter θ (but the superposition of both transformations T and P does not change the sign of θ). In the case of the three-dimensional space ($\theta = 0$ and $\theta = \pi$) under the change of sign of θ, the representation remains unchanged ($e^{i\pi} = e^{-i\pi}$), and in this case, the symmetries T and P are conserved. For the two-dimensional systems, transformation $\theta \to -\theta$ leads to the change of representation (although the superposition of both symmetries is conserved).

3.2.5 *Relation between spin and statistics*

Although the Pauli theorem about the relationship between spin and statistics is associated with the relativistic field theory, the non-relativistic quantum mechanics also contains some hints pertaining to the relationship between quantum statistics of particles and their spin. In the case of the three-dimensional Euclidean space, the rotation group becomes a three-dimensional orthogonal group $\mathcal{O}(3)$.[9]It is a doubly connected group[10] and some of its representations can be multi-valued.[11] To obtain a single-valued representation, one should consider a universal covering group of group $\mathcal{O}(3)$, that is, $\mathcal{SU}(2)$,[12] which is already simple-connected. Representations of this group divide into two classes. The first class includes representations for which a rotation at 2π constitutes the elementary element, therefore, they constitute standard rotations. In the second class the elementary element is the rotation at 4π. As there are no other covering groups for the group $\mathcal{O}(3)$ than $\mathcal{SU}(2)$ (and the latter is doubly-covering), only two allowed rotation symmetries of the wave function exist: standard rotations associated with the integral spin (angular momentum) and non-standard rotations associated with the half-integral spin.

In the case of a system of particles on the Euclidean plane R^2, there is only one axis of rotation. This makes the rotation group $\mathcal{O}(2)$ for this space Abelian. This group is isomorphic with the one-dimensional infinite-connected group of unitary matrices. Group $\mathcal{U}(1)$ has infinitely many coverings and $\mathcal{R}$ (the set of integers with summation as the group operation) constitutes its universal covering. If $\mathcal{Z}_{(x)}$ denotes a group generated by number x ($\mathcal{Z}_x = \{a : a = n \cdot x, n \in \mathcal{Z}\}$),

[9]$\mathcal{O}(3)$ is the set of all real unitary matrices 3×3 with determinant equal $+1$.

[10]As each rotation is attributed to a vector of length equal to the value of angle of rotation and direction consistent with the axis of rotation, all elements of the group $\mathcal{O}(3)$ fill up a sphere of radius π, and points that are opposite on the sphere (e.g. rotation on axis $\hat{z}$ at angles π and $-\pi$) are treated as identical. Points belonging to the sphere constructed in such a way form a double-connected space: all curves connecting pairs of points K_1 and K_2 of this sphere divide into two classes of homotopies. One class is represented by a curve connecting points K_1 and K_2 and consisting of points located in the distance shorter than π from the sphere centre; the other class is represented by a curve connecting points K_1 and K_2 including some points from the surface of the sphere.

[11]For particles with half-integral spin, a rotation of the system at angle 2π (a neutral element of group $\mathcal{O}(3)$) around any axis, a scalar wave function of particles changes its sign, that is, the wave function phase changes by π.

[12]Group $SU(2)$ is a unitary unimodular group of matrices 2×2, that is, it is a group of all unitary matrices 2×2 with determinant equal $+1$.

then $\mathcal{O}(2)$ becomes a factor group $\mathcal{R}/\mathcal{Z}_{(2\pi)}$ ($\mathcal{Z}_{(2\pi)}$ is the normal subgroup of $\mathcal{R}$). The choice of groups $\mathcal{Z}_{(2\pi k)}$, where $k \in R$, determines all coverings of the group $\mathcal{O}(2)$, and all representations of the group $\mathcal{R}$, in the form: $\xi(x) = e^{i\alpha/k}$, where α is an element of the group $\mathcal{R}$ (representations of $\mathcal{R}$ are isomorphic with the factor groups $\mathcal{R}/\mathcal{Z}_{2\pi k}$). These representations can be described by a parameter γ defined as the operation of rotation by angle 2π acting on the wave function:

$$R_{2\pi}\psi = e^{i\gamma}\psi, \tag{3.3}$$

where $\gamma = 2\pi/k$. The operation of rotation by an arbitrary angle α acting on a wave function can be thus written as follows

$$R_\alpha\psi = e^{i(\gamma\alpha/2\pi)}\psi. \tag{3.4}$$

The expression of the rotation by an angle α in terms of a spin operator

$$R_\alpha = e^{i\alpha s/\hbar}, \tag{3.5}$$

leads to the following expression for the spin:

$$s = \frac{\hbar\gamma}{2\pi}. \tag{3.6}$$

There is no limitation imposed on the value of the parameter γ, therefore, the particles with spin different than integer or half-integer can also be realized on the plane.

Considering a pair of identical particles and a rotation by an angle π around an axis passing by the centre of mass of the particles (perpendicular to the line linking the particles), one can notice that this operation is identical to the exchange of particle positions. Taking into account that the representations of the rotation group $\mathcal{O}(2)$ and braid group B_N are expressed through the constant parameters, the spin γ and the statistics type θ, respectively, the relationship between spin and statistics of two-dimensional particles can be established (Balachandran *et al.* 1990): $s = \hbar\theta/2\pi$.

3.2.6 *Aharonov–Bohm effect*

The braid group formalism was introduced at the end of 1960s by Schulman (1968) and Laidlaw and DeWitt (1971). However, they focused their interest on the three-dimensional systems and, therefore, their fundamental group of the configuration space did not lead to fractional statistics. In 1976, Leinaas and Myrheim (1977) proved that the non-trivial topology of the configuration space of two-dimensional systems given by the first homotopy group can result in the appearance of particles of neither boson nor fermion quantum statistics. Only at the beginning of 1980s did the field theory considerations about non-trivial $2+1$ realizations of quantum electrodynamics (dyons and baby-skyrmions) lead to the formulation of the physical model (Wilczek 1982*a*, *b*) referred to the system of particles obeying fractional quantum statistics. In that approach, the problem of particles on the plane was reduced to the collective Aharonov–Bohm effect (Aharonov and Bohm 1959; Peshkin 1981). In such systems, particles move around an area where magnetic flux density equals zero. However, the vector potential in this area is different from zero and it is manifested in a specific particle interaction. Let us consider a charged particle moving in a field of the vector potential generated by an infinitely long solenoid of the radius R which is placed along axis $\hat{z}$. The induction $\mathbf{B}$ vanishes outside the solenoid and is $\mathbf{B} = B\hat{n}_z$ inside. The vector potential outside the solenoid is given by

$$A_r = A_z = 0, \quad A_\varphi = \frac{\Phi}{2\pi r}, \tag{3.7}$$

where Φ is the magnetic field flux inside the solenoid.

Outside the solenoid one gets $\mathbf{B} = \mathrm{rot}\,\mathbf{A} = 0$, so it can be assumed that $\mathbf{A} = \nabla\chi$, where the gauge function χ equals:

$$\chi = \frac{\Phi}{2\pi}\varphi. \tag{3.8}$$

It is not a single-valued function of angle φ and it increases by Φ when $\varphi \to \varphi + 2\pi$.[13] It is because the relevant configuration space is multiply connected and the gauge function is multiple valued that an infinite singular line appears here. This singularity can be characterized by the winding number.[14] The winding number describes a point singularity (generally, of an order parameter) in the two-dimensional space, therefore an infinite string singularity in three-dimensional space is topologically equivalent to a two-dimensional point singularity. Thus, the group $\mathcal{Z}$ constitutes the fundamental group $\pi_1(Q_N(M))$ of the configuration space for a particle in the space R^3 with a string (or, equivalently a particle on the plane with a point singularity).

The presence of the vector potential results in the change of the quantization of the total angular momentum. The periodic boundary conditions connected with the gauge function χ should be imposed on the wave function:

$$\psi(\varphi + 2\pi) = e^{i\Phi q}\psi(\varphi), \tag{3.9}$$

This condition leads to the following eigenfunction of the zth component of the angular momentum:

$$\psi(\varphi) = (\mathrm{const})e^{i(k-(\Phi q/2\pi))\varphi}, \tag{3.10}$$

[13] It is the ambiguity of the gauge which leads to the Aharonov–Bohm effect. If the following equation: $\mathbf{B} = \mathrm{rot}\,\mathbf{A} = \mathrm{rot}\,\mathrm{grad}\,\chi = 0$ would be true everywhere, the magnetic flux would equal zero in all space.

[14] The vector potential outside the solenoid is a continuous function; thus, if the solenoid is encircled by a single loop, the angle at which the vector $\mathbf{A}$ moves around when we move along the loop is $2\pi n$, where n is a natural number. The factor n is called the winding number and it characterizes the type of singularity. As the winding number changes only by one (as it is an integer) and $\mathbf{A}$ is a continuous field outside the solenoid, the winding number does not change if one continuously deformes the loop around the solenoid. In our case, the angle at which $\mathbf{A}$ moves around equals 2π. Such an object is said to have a winding number equal to 1.

where k is an integer. The projection of the angular momentum on the axis z can assume values $l_z = k - (\Phi q/2\pi)$. This result corresponds to the reduction of a three-dimensional system effectively to a two-dimensional one. For this case the rotation group is Abelian and has an infinite number of one-dimensional unitary representations which corresponds to the fact that the fractional angular momenta may appear.

If one considers the system of N indistinguishable particles on the plane, each of them carrying the same flux,[15] each particle generates a vector potential. When particle A describes a loop around particle B, the phase of the many-particle wave function changes by the value $2\theta = 2\Phi q$. The factor 2 is related to the fact that the vector potential from the particle A changes at the position of particle B and such a loop can be represented as a double exchange of positions of the particles A and B. In this case the factor θ is related to quantum statistics and defines the representation of the braid group. The fractional statistics is defined by the order parameter, in this case, in the form of the vector potential. The singularities of the order parameter on a given manifold correspond to the singular points of the configuration space $Q_N(M)$ found when the permutation group S_N acts on $Q_N(M)$ (diagonal points).[16]

3.2.7 *Fractional statistics in one-dimensional system*

In Section 2.2, it was shown that the formalism of quantum statistics based on the notion of the braid group does not correctly describe one-dimensional systems.

In the case of the circle S^1, one cannot isolate a subgroup $\Sigma_N(S^1)$ within the braid group because the positions of the two particles cannot be exchanged when $N > 2$. Moreover, each change of the arrangement of particles results in describing non-contractible loops by particles in the space $Q_N(S^1)$. Thus, the braid group is generated by the operations which are, on the one hand, the generators of $\pi_1(F_N(M))$ (they generate loops, non-contractible in S^1), and on the other hand, the generators of $\Sigma_N(M)$ (exchange particles positions). In the case of two-dimensional and three-dimensional manifolds, the generators from the set $\mathcal{L}$ can be expressed through generators from set $\mathcal{P}$; otherwise the sets $\mathcal{L}$ and $\mathcal{P}$ are separable. In the case of the line R^1, the braid group is denoted by $\pi_1(Q_N(R^1)) = \mathcal{E}$ and in such a case it is pointless to talk about quantum statistics. However, in the 1970s, Leinaas and Myrheim (1977) proved that fractional statistics may exist for the system of two particles on the line. In that case, the

[15]It is not necessary to assume that the flux which is attached to particles is always of a magnetic field nature.

[16]A closed non-contractible path in the space $F_N(M)$ corresponds to the fact that particles have described loops around other particles and that is why the phase of the wave function changes according to the suitable representation of the braid group. On the other hand, this means that on a real manifold, particles have described loops around singularities of the vector potential generated by other particles, which results in same change of the wave function phase.

configuration space is $Q_2(R) = (R^2 \setminus \Delta)/S_2$. If x_1 and x_2 refer to particle positions, then Δ denotes the straight line $x_1 = x_2$ on the plane, and the division by S_2 leads to the points (x_1, x_2) and (x_2, x_1) becoming identical on the plane. Thus, $Q_2(R)$ becomes the half-plane with no edge. The Hamiltonian of two free particles on R is given by:

$$H = -\frac{\hbar^2}{2m}\left(\frac{\partial^2}{\partial x_1^2} + \frac{\partial^2}{\partial x_2^2}\right) = -\frac{\hbar^2}{4m}\frac{\partial^2}{\partial x^2} - \frac{\hbar^2}{m}\frac{\partial^2}{\partial z^2}, \tag{3.11}$$

where $x = (x_1 + x_2)/2$, $z = |x_1 - x_2|$. The wave function disappears at the boundary (what is usually assumed for such cases) giving fermion wave functions. It is suggested (Leinaas and Myrheim 1977) that a weaker boundary condition should be assumed:

$$\left[\psi^*(x,z)\frac{\partial\psi(x,z)}{\partial z} - \frac{\partial\psi^*(x,z)}{\partial z}\psi(x,z)\right]_{z=0} = 0 \tag{3.12}$$

for each x (disappearance of the normal component of a probability current). The above equation leads to a condition:

$$\left[\frac{\partial\psi(x,z)}{\partial z}\right]_{z=0} = \eta\psi(x,0), \tag{3.13}$$

where η is a real parameter. The value $\eta = 0$ corresponds to bosons (then $\psi(x,0)$ can be different than 0) and $\eta^{-1} = 0$ corresponds to fermions (then $\psi(x,0) = 0$). Intermediate values correspond to neither bosons nor fermions.

It should be said that for the one-dimensional case, the set of diagonal points is the boundary of the configuration space and that is why different boundary conditions can lead to different statistics. In higher dimensional spaces—the set of diagonal points is embedded in corresponding configuration spaces. For example, if one considers two particles on the plane, then their configuration space is equivalent to a cone of angle 60° without the cone vertex (it is a diagonal point). Therefore, the space $Q_2(R^2)$ is infinitely connected[17] and its topological properties are characterized by the winding number around the cone vertex which corresponds to the braid group structure and leads to the fractional statistics ($B_2 = \mathcal{Z}$). For two particles on the straight line, their configuration space is simple-connected. In the case of the space R^3 for two particles, the configuration space is doubly connected ($\pi_1(Q_2(R^3)) = S_2 = \mathcal{Z}_2$).

3.3 Non-Abelian statistics

In analogy with the idea of parastatistics (which corresponds to higher dimensional representations of the symmetric group) the non-Abelian braid statistics can be found when higher dimensional (non-Abelian) irreducible representations of the braid group are considered (Moore and Read 1991; Fradkin *et al.* 1998;

[17]That is, the first homotopy group of this space has an infinite number of elements.

Isakov *et al.* 1999; Polychronakos 2000; Ivanov 2001). The example of the non-Abelian braid statistics was found by Moore and Read for quasiholes in the quantum Hall Pfaffian state (Moore and Read 1991; Nayak and Wilczek 1996; Fradkin *et al.* 1998; Read and Rezayi 1999).

The non-Abelian statistics of quasiholes in the Pfaffian state was further analysed by Nayak and Wilczek (1996) who showed that interchanges of quasiholes in the $2N$ quasihole state are described by the group $SO(2N) \times U(1)$ (which contains the braid group as a subgroup). A single interchange of two quasiholes can be interpreted as a local π rotation in the real space (around the centre of mass of these two quasiparticles; positions of other quasiparticles remain unchanged). Such an interchange is given by the element (related to $\pi/2$ rotations) of the spinorial representation of the group $SO(2N) \times U(1)$. Hence, the group $SO(2N) \times U(1)$ can be regarded as a continuous extension of the braid group B_N. Below, we present the different concepts (given by Wilczek (1998)) of non-Abelian statistics based on the projective representation of the permutation group (and hence, valid in any dimension).

3.3.1 *Projective permutation statistics*

Wilczek (1998) proposed a new way of defining non-Abelian statistics—projection permutation statistics. He was motivated by the correspondence between projective representations of the permutation group S_N and spinor representations of the group $SO(N)$ (dimensions of representations are the same). The new statistics is independent of a dimension of the space.

In the case of ordinary linear representations of the group G, an element $g \in G$ is associated with a linear operator $A(g)$ (which acts on vectors in the Hilbert space) and the following relation is satisfied:

$$g_1 g_2 = g_3 \implies A(g_1)A(g_2) = A(g_3). \tag{3.14}$$

However, pure quantum states are described up to the phase of the wave function, by a ray $e^{i\phi}\Psi$ (ϕ is an arbitrary factor from $[0, 2\pi)$), not by the vector Ψ. Then one can say that the operators A transform one ray into another and a representation of the symmetry group can be regardered as acting on the rays in the Hilbert space. If one replaces the relation (3.14) with the following:

$$g_1 g_2 = g_3 \implies A(g_1)A(g_2) = e^{i\phi(g_1,g_2)}A(g_3), \tag{3.15}$$

a projective representation is obtained (Hamermesh 1962).

The projective representation $\tilde{S}_N$ of the symmetric group S_N can be given as a vector representation of the central extension of the group S_N by $U(1)$ (it is a covering group of S_N). For $N \geq 4$, the only extensions $\tilde{S}_N$ that cannot be reduced to the linear representation are extensions of the group S_N by the group Z_2 (Read 2002). The group $\tilde{S}_N$ is defined by the generators σ_i ($i = 1, 2, ...,$ $N-1$, compare with eqns (2.14) and (2.15)) and the additional generator z. The relations defining the group $\tilde{S}_N$ are as follows (Hamermesh 1962):

$$z^2 = e, \tag{3.16}$$

$$z\sigma_i = \sigma_i z, \tag{3.17}$$

$$\sigma_i^2 = z, \tag{3.18}$$

$$\sigma_i\sigma_k\sigma_i = \sigma_k\sigma_i\sigma_k, \tag{3.19}$$

$$\sigma_i\sigma_j = z\sigma_j\sigma_i, \tag{3.20}$$

where $i = 1, 2, \ldots, N-1$, $|i-j| > 1$, $|i-k| = 1$. The generator z is the central element of the group (it commutes with all others generators of $\tilde{S}_N$ (3.17)). One can see from eqn (3.16) that in the linear representation $\mathcal{R}(\tilde{S}_N)$ the image of the generator z is given by $\mathcal{R}(z) = \pm 1$. If $\mathcal{R}(z) = 1$, then the representation $\mathcal{R}(\tilde{S}_N)$ is the vector representation of the permutation group S_N. If $\mathcal{R}(z) = -1$, then the linear representation $\mathcal{R}(\tilde{S}_N)$ becomes the projective representation of the permutation group S_N. It follows eqn (3.18) that $\sigma_i^2 = z \neq e$—a double exchange of two neighbouring particles is not related to the neutral element of the group as in the braid group (however, $\sigma_i^4 = e$), not as in the symmetric group. The generator z appears also in the relation (3.20), and hence, contrary to the braid group, operations of an exchange of particles in two well-separated pairs do not commute. It should be noted, however, that in the case of projective non-Abelian statistics, some problems with the quantum locality arise (Read 2002).

One should stress again, that, in contrast to the braid group, the projective permutation statistics does not depend on a dimension of the space (Read 2002).

4

TOPOLOGICAL APPROACH TO COMPOSITE PARTICLES IN TWO DIMENSIONS

4.1 Mathematical model of composite particles

In this chapter we present the generalization of the braid group formalism (Wieczorek and Jacak 2002) for composite particles on the Euclidean plane R^2. Let us consider a change of the phase $\Delta\phi_0$ of the wave function when two neighbouring particles (with indices i and $i+1$) exchange their positions. For anyons, $\Delta\phi_0$ is equivalent with the statistical parameter θ—the one-dimensional representation of the braid group. For composite fermions, the factor $\Delta\phi_0$ equals $(2p+1)\pi$. The quantum statistical properties give the value of $\theta_{\mathrm{cf}} = (2p+1)\pi$ for composite fermions, where p is an integer (and $\theta_{\mathrm{cb}} = 2p\pi$ for composite bosons). The possible values of the parameter θ for anyons and composite particles on R^2 are shown in Fig. 4.1. The phase factor $e^{i\theta}$ is, however, a periodic function of the parameter θ, hence its value is equal for fermions and for composite fermions (analogically for bosons and composite bosons), but the values θ_{cf} and θ_{f} are not equal. Both parameters θ_{cf} and θ_{cb} do not belong to the interval $(-\pi, \pi]$ of permissible values of the parameter θ, obtained within the braid group formalism. It indicates that the braid group formalism is not appropriate for the composite particles on a plane R^2.

The physical model of composite fermions can be based on the generalization of the idea of transmutation of statistics, that is, on attaching a flux to the charged particles. In particular, two-dimensional charged fermions can be defined as bosons of a charge e carrying a point flux hc/e (the flux quantum); composite fermions are obtained from fermions of a charge e by attachment of $2p$ flux quanta (or from bosons by attachment of $2p+1$ flux quanta). The mathematical description of composite fermions requires the translation of the Aharonov–Bohm effect into the mathematical language. The Aharonov–Bohm effect consists of a change of a phase of the wave function (which is periodic with period 2π) when a charged particle describes a loop around a magnetic flux (vortex-line). For a single exchange of particle positions, the change of the

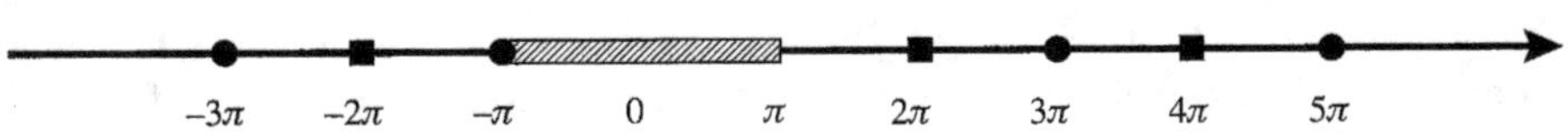

FIG. 4.1. Statistical parameter θ for a system of particles on the Euclidean plane R^2; anyons (dashed area), composite fermions (•) and composite bosons (■).

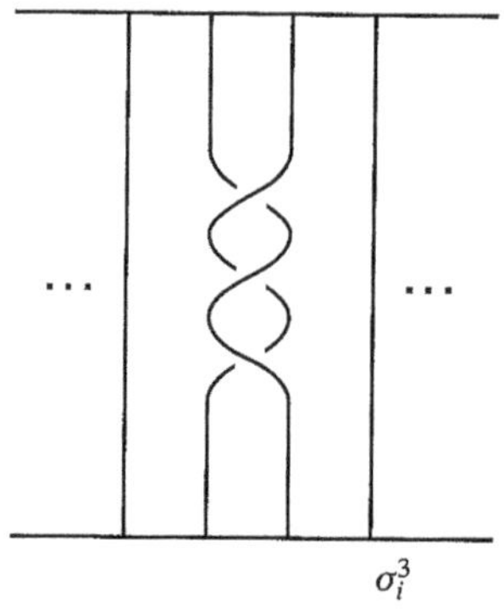

FIG. 4.2. Geometric braid σ_i^3.

phase is not greater than π; however, even a larger change can occur when the trajectories of interchanging of particles are restricted. Let us note that the n-fold interchange of two fermions (with indices i and $(i+1)$) (the element σ_i^n of the braid group B_N, see Fig. 4.2) results in a change of the phase of the wave function by $n\pi$—the same change as for a single exchange of composite fermions with $p = (n-1)/2$. If one forbids the interchanges of these two particles, corresponding to the elements $\sigma_i, \sigma_i^2, \ldots, \sigma_i^{n-1}$, then one gets the exchanges of composite particles (σ_i^n only). However, the elements σ_i^n do not describe the interchanges of the non-neighbouring particles, for example, particles with indices i and j ($j \geq i+2$) can exchange only by the multiple operations σ_k^n. It should be taken into account that the interchange of particles ith and jth is obtained in three steps: (a) the particle i is moved to the position $j-1$ (element α of the braid group), (b) they interchange (element σ_{j-1}^n of braid group), (c) finally the particle in the position $j-1$ returns to the position i (element α^{-1} of the braid group) (see Fig. 4.3). Then the single interchange of two arbitrary particles is given by the element:

$$\omega_{\alpha,\sigma_i} = \alpha\sigma_i^n\alpha^{-1} \tag{4.1}$$

of the braid group B_N, where $\alpha \in B_N$. We will call particles exchanging in this way *the looped particles of the nth order.* A single interchange of these two particles involves $n-1$ additional closed non-contractable paths in the configuration space $Q_N(R^2)$ ($p = (n-1)/2$ closed paths in the real space). If n is an even number, then the elements $\alpha\sigma_i^n\alpha^{-1}$ do not describe the exchange of particle positions; however, they generate a subgroup of the pure braid group. Therefore, the number n should be odd.

Let us denote by Ω_N^n the group describing all exchanges of looped particles, generated by the elements ω_{α,σ_i}. The group Ω_N^n is a normal subgroup of the

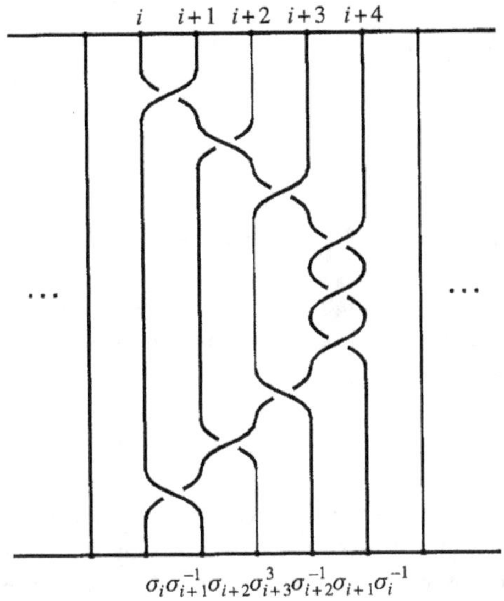

FIG. 4.3. Geometric braid $\sigma_i\sigma_{i+1}^{-1}\sigma_{i+2}\sigma_{i+3}^{3}\sigma_{i+2}^{-1}\sigma_{i+1}\sigma_i^{-1}$ which describes single interchanging of two non-neighbouring looped particles with $n = 3$.

braid group B_N due to the form of the generators ω_{α,σ_i} (see Eqn (4.1)). For the few simplest cases, the explicit form of the group Ω_N^n can be found. In order to determine the group Ω_N^n, the factor group has to be found.

4.1.1 *Factor groups B_N/Ω_N^n*

The factor group B_N/Ω_N^n is generated by the generators σ_i which satisfy the relations (2.14), (2.15) (defining the generators of the braid group B_N), and the relations

$$\sigma_i^n = e, \tag{4.2}$$

where $N - 1 \geq n \geq 1$.

1. $N = 2$: For $N = 2$ the braid group B_2 is the cyclic group $\mathcal{Z}$ of an infinite order which is generated by σ. Therefore, the factor group B_2/Ω_2^n is the cyclic group $\mathcal{Z}_n$ (generated by σ) of the rank n.
2. $N = 3$: For $N = 3$ the factor group B_N/Ω_N^n is generated by the two generators: σ_1 and σ_2 which satisfy the relations

$$\sigma_1\sigma_2\sigma_1 = \sigma_2\sigma_1\sigma_2, \tag{4.3}$$

$$\sigma_1^n = \sigma_2^n = e. \tag{4.4}$$

Another presentation of the factor group B_N/Ω_N^n can be found when introducing the new generators: $A = \sigma_1\sigma_2$ and $B = \sigma_1^{-1}\sigma_2^{-1}\sigma_1^{-1}$ which satisfy the relations

$$B^{-2} = A^3, \tag{4.5}$$

$$(VA)^n = e. \tag{4.6}$$

Therefore, B_N/Ω_N^n is the group $< -2, 3|n >$ (Coxeter and Moser 1984).

For $n = 3$ $< -2, 3|3 >=< 2, 3, 3 >$, where $< 2, 3, 3 >$ is the binary tetrahedral group of the rank 24.

For $n = 5$ $< -2, 3|5 >=< 2, 3, 5 > \otimes \mathcal{Z}_5$, where $< 2, 3, 5 >$ is the binary icosahedral group of the rank 120. Then, B_3/Ω_3^5 is the group of the rank $120 \cdot 5 = 600$.

For $n \geq 6$, and also for all odd $n > 5$, every groups $<-2, 3|n>$ are the groups of an infinite order.

4.1.2 *Group* Ω_N^n

1. $N = 2$: For a system of two particles on the Euclidean plane R^2, the braid group B_2 is the cyclic group $\mathcal{Z}$ (generated by σ) of an infinite order. The group Ω_2^n is also the cyclic group $\mathcal{Z}$; however, it is generated by σ^3.
2. $N = 3$, $n = 3$: The presentation of the group Ω_3^3 can be found using the Reidemeister–Schreier method (Coxeter and Moser 1984). The group is generated by the four generators:

$$\begin{aligned}
\omega_a &= \omega_{e,\sigma_2} = \sigma_2^3, \\
\omega_b &= \omega_{\sigma_1,\sigma_2} = \sigma_1 \sigma_2^3 \sigma_1^{-1}, \\
\omega_c &= \omega_{e,\sigma_1} = \sigma_1^3, \\
\omega_d &= \omega_{\sigma_1^{-1},\sigma_2} = \sigma_1^{-1} \sigma_2^3 \sigma_1 = \sigma_2 \sigma_1^3 \sigma_2^{-1},
\end{aligned} \tag{4.7}$$

which satisfy the relations:

$$\omega_a \omega_b \omega_c \omega_d = \omega_b \omega_c \omega_d \omega_a = \omega_c \omega_d \omega_a \omega_b = \omega_d \omega_a \omega_b \omega_c. \tag{4.8}$$

The geometric braids related to generators are shown in the Fig. 4.4. The elements $\omega_a \omega_b \omega_c \omega_d$, $\omega_b \omega_c \omega_d \omega_a$, $\omega_c \omega_d \omega_a \omega_b$, and $\omega_d \omega_a \omega_b \omega_c$, which occur in the Eqn (4.8), represent the pure braids of B_3 and these operations do not change the initial order of the particles.

The group Ω_3^3 is generated not only by the operation of the exchange of two neighbouring particles: ω_a and ω_c, but also by the two inequivalent operations of the interchange of the first and the third particle: ω_b and ω_d. The generators ω_b and ω_d cannot be written in terms of $\{\omega_a, \omega_c\}$, that is, the group Ω_3^3 cannot be generated by an interchange of neighbouring particles.

1. $N = 3$, $n = 5$: The factor group B_3/Ω_3^5 is $< 2, 3, 5 > \otimes \mathcal{Z}_5$, where $< 2, 3, 5 >$ is the binary icosahedral group (Coxeter and Moser 1984) of the rank 120, and $\mathcal{Z}_5$ is the cyclic group of the rank 5. Then B_3/Ω_3^5 is the group of the rank 600 and the group Ω_3^5 has a finite presentation.

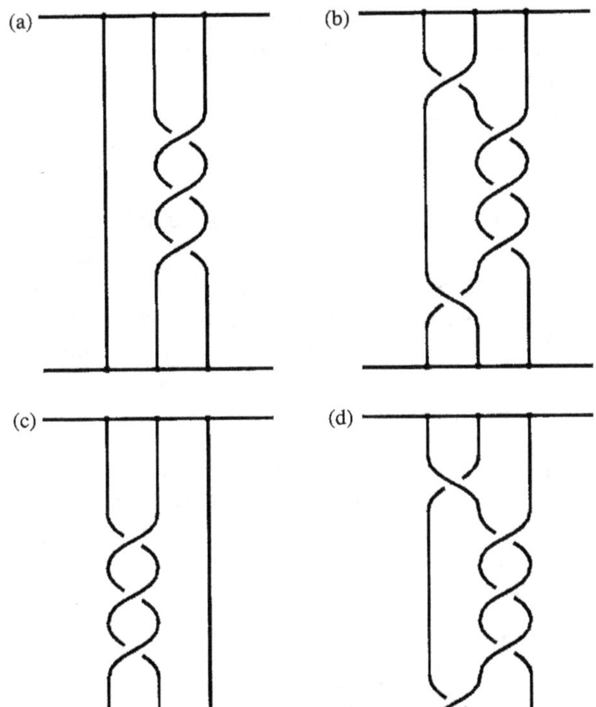

FIG. 4.4. Geometric braids of the generators (a) $\omega_a = \sigma_2^3$, (b) $\omega_b = \sigma_1\sigma_2^3\sigma_1^{-1}$, (c) $\omega_c = \sigma_1^3$, and (d) $\omega_d = \sigma_1^{-1}\sigma_2^3\sigma_1$, of the group Ω_3^3.

For $n = 5$ and $N > 3$, as well as for $n \geq 5$ and $N > 2$, the factor group B_N/Ω_n^N becomes a group of an infinite order (Coxeter and Moser 1984).

For the system of indistinguishable particles in the Euclidean three-dimensional space R^3, the fundamental group $\pi_1(Q_N(R^3)) = \mathcal{S}_N$ is the permutation group, and every even permutation is equivalent to the unit element of the group. Then if n is an odd number, $\sigma_i^n = \sigma_i^{n-1}\sigma_i = \sigma_i$ so that $\Omega_N^n = \mathcal{S}_N$, and looped particles cannot appear in the system of indistinguishable particles in the Euclidean three-dimensional space.

4.1.3 *One-dimensional unitary representations of the group* Ω_N^n

The generators of the group Ω_N^n are of the form $\omega_{\alpha,\sigma_i} = \alpha\sigma_i^n\alpha^{-1}$, $\alpha \in B_N$ (by analogy with Eqn (4.1)) and then

$$[\Omega_N^n]_{ab} = \bigotimes_{i=1}^{N-1} \mathcal{Z}_i, \tag{4.9}$$

where $\mathcal{Z}_i$ is the cyclic group (generated by σ_i^n, $1 \leq i \leq N-1$) of an infinite order. The group $[\Omega_N^n]_{ab}$ is the free Abelian group generated by $N-1$ generators.[1] Every one-dimensional unitary representation $\mathcal{W}_{\tilde{\theta}_1}(\Omega_N^n)$ described by the same statistical parameter $\tilde{\theta}_1 \in (-\pi, \pi]$, can be determined by the representation of the braid group B_N:

$$e^{\tilde{\theta}_1 i} = \mathcal{W}_\theta(\omega_{\alpha\sigma_i}) = \mathcal{W}_\theta(\sigma_i^n) = [\mathcal{W}(\sigma_i)]^n = e^{n\theta i} \tag{4.10}$$

where $\theta, \tilde{\theta}_1 \in (-\pi, \pi]$. The factor $e^{\tilde{\theta}_1 i}$ describes the effective change of the wave function phase due to a single interchange of two looped particles of the nth order. Let us introduce a new statistical parameter θ_1 for the looped particles of the nth order:

$$\theta_1 = n\theta. \tag{4.11}$$

The parameter θ_1 describes the actual change of the wave function phase for a single interchange of two looped particles. Although $e^{\theta_1 i} = e^{\theta i}$, the distinction between θ_1 and θ shows the difference between looped particles and non-looped particles. This difference results from the limitation of the ways of interchanging of particle positions. Then the actual change of the wave function phase due to a single interchange of particle positions is the same for composite fermions and looped particles when $\theta_{\text{cf}} = \theta_1$.

The wave function proposed by Greiter and Wilczek (1990) for composite fermions, based on the idea of attaching point fluxes to ordinary two-dimensional fermions, can be written in the form (in bosonic representation):

$$\Psi = \prod_{i<j} \left[\frac{(z_i - z_j)^{(2p+1)/n}}{|z_i - z_j|^{(2p+1)/n}} \right]^n \chi_{\text{b}} = \left[\prod_{i<j} \frac{(z_i - z_j)^{(2p+1)/n}}{|z_i - z_j|^{(2p+1)/n}} \right]^n \chi_{\text{b}}, \tag{4.12}$$

where χ_{b} is the boson wave function, p is an integer, n is an odd number which corresponds to the order of the looped particles. The factor

$$\frac{(z_i - z_j)^{(2p+1)/n}}{|z_i - z_j|^{(2p+1)/n}}$$

represents a change of the wave function phase in the system for a single interchange of particle positions two (particle indices are i and j, and statistics parameter $\theta = ((2p+1)\pi)/n$). The power n in Eqn (4.12) corresponds to n-fold exchange of particle position.

Note that the composite fermions with $\theta_{\text{cf}} = (2p+1)\pi$ do not have a unique description in the model. They can be considered as anyons with $\theta = \theta_{\text{cf}}/n = ((2p+1)\pi)/n$ (where n is an odd number), when their interchange is given by σ_i^n

[1] The free Abelian group is an Abelian group generated by finite generators with no additional defining relations.

and thus there are infinite number of ways (one for each n) to define composite particles.

4.2 Configuration space for the system of composite particles

The homotopy group (which describes interchanges of non-looped particles) transforms into the group Ω_N^n for looped particles upon the restriction introduced in Section 4.1 (imposed on an interchange manner of particle positions). Thus, looped particles can be represented as non-looped particles with the new configuration space. Let us assume that the configuration space of the system of N looped particles of the nth order, Γ_N^n, has the fundamental group Ω_N^n. The forbidden interchanges of particle positions and the corresponding elements of the braid group $B_N(R^2)$ are described by paths which are open in the new configuration space Γ_N^n (they are not related to exchange of the looped particles positions). Thus, the space Γ_N^n happens to be the covering space for $Q_N(R^2)$.

Within the covering projection, the inverse images of the closed paths in the base space may be the open paths in the covering space. Thus, some paths which are closed in Q_N would be open after lifting to the space Γ_N^n. The configuration space Q_N is a manifold so that there exists the covering space, with fundamental group Ω_N^n, for every subgroup $\Omega_N^n \subset B_N$. Therefore, the space Γ_N^n is the covering space for $Q_N(R^2)$. The group Ω_N^n is a normal subgroup of B_N and (up to an isomorphism) there is only one covering space Γ_N^n with the characteristic group Ω_N^n of covering projection and the cover is regular. In the case of regular covers, the factor group B_N/Ω_N^n acts on a fibre in a transitive way, that is, each orbit of this group is a full fibre and the multiplication factor of the covering projection is $|B_N/\Omega_N^n|$. The closed path in $Q_N(R^2)$, given by the element $g \notin \Omega_N^n$ (e.g. σ_i, σ_i^2), becomes open when it is lifted to the space Γ_N^n—it is the path connecting different points of the fibre. If a closed path in $Q_N(R^2)$, given by element $g \in \Omega_N^n$, is lifted, then the closed path in the space Γ_N^n is obtained.

4.2.1 *Configuration space for two looped particles*

In the case of a system of two particles on the Euclidean plane R^2, the braid group B_2 is the cyclic group $\mathcal{Z}$ (generated by σ) of an infinite order. The group Ω_2^n is the group $\mathcal{Z}$, but it is generated by σ^3. The factor group B_2/Ω_2^3 is $\mathcal{Z}_3$ and the fibre of the covering projection, which we are seeking, is a space that consists of three points. Leinaas and Myrheim (1977) found a configuration space for a system of two particles on the Euclidean plane R^2 of the form

$$Q_2(R^2) = R^2 \times \mathcal{R}(2,2),$$

where the Euclidean plane R^2 represents the position of the centre of mass and $\mathcal{R}(2,2)$ is the space of two degrees of freedom of the relative motion of the particles. The space $\mathcal{R}(2,2)$ is equivalent to the plane R^2 where opposite points

(a)

(b)

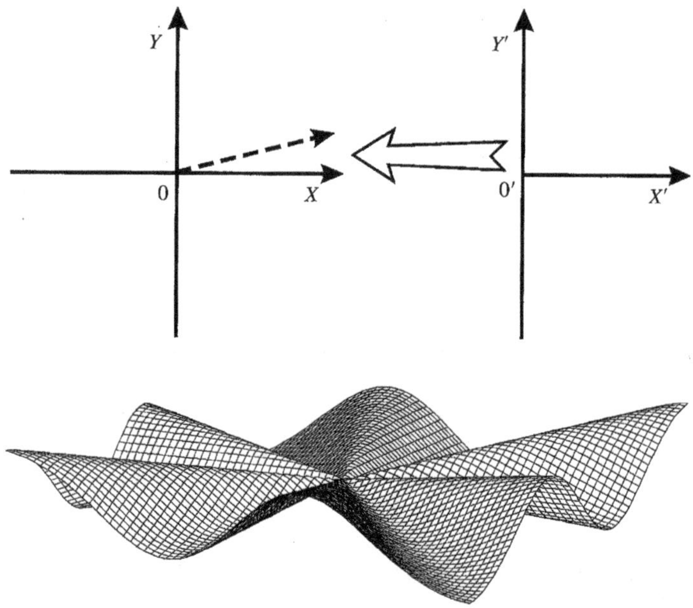

FIG. 4.5. Construction of the relative space $\mathcal{R}_3(2,2)$: (a) construction of the relative space $\mathcal{R}_3(2,2)$ and (b) relative space $\mathcal{R}_3(2,2)$.

are considered identical, excluding the origin of the system of coordinates. It is a cone of the angle 60° with a singular vertex. The fundamental group of the plane R^2 is a trivial one and $\pi_1(Q_2(R^2)) = \pi_1(\mathcal{R}(2,2))$ (Spanier 1966), therefore, a search for a cover of the configuration space $Q_2(R^2)$ is equivalent to a search for the cover of the relative space $\mathcal{R}(2,2)$. The relative space $\mathcal{R}_3(2,2)$ for two looped particles with $n = 3$ can be constructed by the dissection of the plane R^2, which has a singular point in the origin, along one of the semi-axes (e.g. the OX axis) and by an insert of a half-plane in that place (e.g. half-plane $x > 0$) (see Fig. 4.5). This space can alternatively be obtained by sticking together three half-planes. Three points in the space $\mathcal{R}_3(2,2)$ correspond to a single point in the relative space $\mathcal{R}(2,2)$. The lifted paths in space $\mathcal{R}_3(2,2)$ which are obtained from the paths in the space $\mathcal{R}(2,2)$, given by the elements σ and σ^2 of the braid group B_2, are open. Only the lifts of σ^3 paths (and powers of σ^3) are closed paths in the space $\mathcal{R}_3(2,2)$—they describe closed paths around a singular point at the origin. Therefore, the configuration space of a system of looped particles $\Gamma_2^3(R^2)$, which covers the space $Q_2(R^2)$ three times, is

$$\Gamma_2^3(R^2) = R^2 \times \mathcal{R}_3(2,2). \tag{4.13}$$

In the case of a system of two looped particles of an arbitrary order, we have $B_2/\Omega_2^n = \mathcal{Z}_n$ and the covering fibre is a discrete space of n points. The relative space $\mathcal{R}_n(2,2)$ consists of n half-planes (with a singular point in the origin) stuck together and then a point in the space $\mathcal{R}(2,2)$ corresponds to n points in the space $\mathcal{R}_n(2,2)$. The lifting of powers of σ^n paths (σ^n and its powers are the elements of the group $\pi_1(Q_2(r^2))$) to the covering space $\mathcal{R}_n(2,2)$ is represented by closed paths surrounding a singular point. Hence the space $\Gamma_2^n(R^2)$ which covers the space $Q_N(R^2)$ n times is given by

$$\Gamma_2^n(R^2) = R^2 \times \mathcal{R}_n(2,2). \tag{4.14}$$

4.2.2 *Configuration space of the system of three looped particles of the third order*

The configuration space of the system of three particles in the Euclidean plane R^2 is given by Eqn (2.1):

$$Q_3(R^2) = ((R^2)^3 \setminus \Delta)/\mathcal{S}_3. \tag{4.15}$$

The position of the centre of mass is invariant under $\mathcal{S}_3$ and the configuration space can then be rewritten in the form:

$$Q_3(R^2) = R^2 \times ((R^2)^2 \setminus \widetilde{\Delta})/\mathcal{S}_3, \tag{4.16}$$

where the first factor R^2 represents the position of the centre of mass of the system and $\widetilde{\Delta}$ is a collection of three hyperplanes corresponding to Δ. The space $((R^2)^2 \setminus \widetilde{\Delta})/\mathcal{S}_3$ can be projected on the three-dimensional sphere S^3 and therefore

$$((R^2)^2 \setminus \widetilde{\Delta})/\mathcal{S}_3 = R_+ \times (S^3 \setminus 3S^1)/\mathcal{S}_3, \tag{4.17}$$

where R_+ is a set of positive real numbers and three one-dimensional spheres S^1 are the intersections of the hyperplanes $\widetilde{\Delta}$ with the three-dimensional sphere S^3. The orbit space $(S^3 \setminus 3S^1)/\mathcal{S}_3$ can be obtained by the parametrization of the space $Q_3(R^2)$ by symmetric polynomials and is given by:

$$(S^3 \setminus 3S^1)/\mathcal{S}_3 = S^3 \setminus \mathcal{T}, \tag{4.18}$$

where $\mathcal{T}$ is a torus knot of the type (2, 3) called a trefl (see Fig. 4.6). Then the configuration space $Q_3(R^2)$ is

$$Q_3(R^2) = R^2 \times R_+ \times (S^3 \setminus \mathcal{T}), \tag{4.19}$$

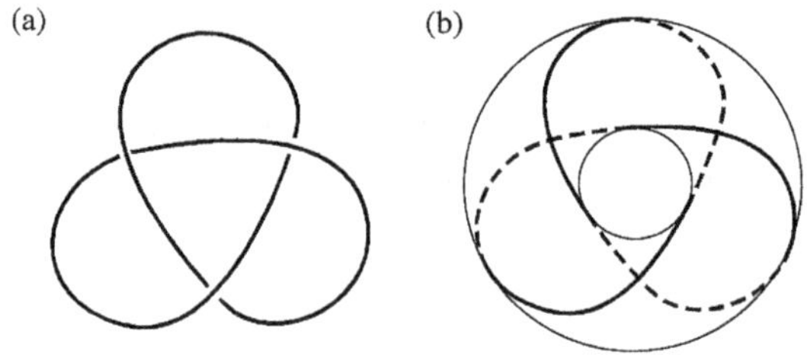

FIG. 4.6. Geometric realization of: (a) a knot $\mathcal{T}$ of a trefl type and (b) a knot $\mathcal{T}$ winded on the surface of the torus.

where $R_+ \times (S^3 \setminus \mathcal{T})$ is the relative space $\mathcal{R}_3(2,3)$. One can note that the Wirtinger presentation of the fundamental group of trefl $\mathcal{T}$ (Rolfsen 1976; Duda 1991) presented in Fig. 4.6 is identical to the braid group B_3 (relations (2.14) and (2.15)).

In order to find a covering space of the space $Q_3(R^2)$, for the characteristic group B_3/Ω_3^3 of the covering projection, the Burau representation (Birman 1974) of the factor group B_3/Ω_3^3 has to be determined first. The Burau representation of the braid group B_3 (Birman 1974; Jones 1987) (after the reduction to the 2×2 representation) is given by

$$\mathcal{B}(\sigma_1) = \begin{bmatrix} -t & 1 \\ 0 & 1 \end{bmatrix}, \quad \mathcal{B}(\sigma_2) = \begin{bmatrix} 1 & 0 \\ t & -t \end{bmatrix}, \tag{4.20}$$

where t is a complex number. In order to find the Burau representation of factor group $\mathcal{B}(B_4/\Omega_4^3)$, the relation (4.2) must be taken into account. Hence, t satisfies the relations

$$\begin{bmatrix} -t^3 & t^2 - t + 1 \\ 0 & 1 \end{bmatrix} = \begin{bmatrix} 1 & 0 \\ 0 & 1 \end{bmatrix}, \tag{4.21}$$

$$\begin{bmatrix} 1 & 0 \\ t^3 - t^2 + t & -t^3 \end{bmatrix} = \begin{bmatrix} 1 & 0 \\ 0 & 1 \end{bmatrix}. \tag{4.22}$$

The above relations are fulfilled if $t = \frac{1}{2}(1 \pm i\sqrt{3})$. Note that the representation $\mathcal{B}(B_3/\Omega_3^3)$ is faithful.[2] The elements of the representation $\mathcal{B}(B_3/\Omega_3^3)$ which have

[2]The representation $\mathcal{R}(G)$ of the group G is faithful if $|\mathcal{R}(G)| = |G|$.

eigenvalue 1 and the corresponding eigenvectors are

$$\mathcal{B}(\sigma_1) = \begin{bmatrix} -t & 1 \\ 0 & 1 \end{bmatrix}, \quad \aleph_{\sigma_1} = \begin{bmatrix} -\frac{1}{2}it \\ \frac{\sqrt{3}}{2} \end{bmatrix}, \tag{4.23}$$

$$\mathcal{B}(\sigma_1^2) = \begin{bmatrix} t-1 & 1-t \\ 0 & 1 \end{bmatrix}, \quad \aleph_{\sigma_1^2} = \begin{bmatrix} -\frac{1}{2}it \\ \frac{\sqrt{3}}{2} \end{bmatrix}, \tag{4.24}$$

$$\mathcal{B}(\sigma_2) = \begin{bmatrix} 1 & 0 \\ t & -t \end{bmatrix}, \quad \aleph_{\sigma_2} = \begin{bmatrix} \frac{\sqrt{3}}{2} \\ \frac{1}{2}i(t-1) \end{bmatrix}, \tag{4.25}$$

$$\mathcal{B}(\sigma_2^2) = \begin{bmatrix} t-1 & 1-t \\ 0 & 1 \end{bmatrix}, \quad \aleph_{\sigma_2^2} = \begin{bmatrix} \frac{\sqrt{3}}{2} \\ \frac{1}{2}i(t-1) \end{bmatrix}, \tag{4.26}$$

$$\mathcal{B}(\sigma_1\sigma_2\sigma_1^2) = \begin{bmatrix} 0 & -t \\ -1 & 1-t \end{bmatrix}, \quad \aleph_{\sigma_1\sigma_2\sigma_1^2} = \begin{bmatrix} \frac{\sqrt{2}}{2} \\ \frac{\sqrt{2}}{2}(t-1) \end{bmatrix}, \tag{4.27}$$

$$\mathcal{B}(\sigma_1^2\sigma_2\sigma_1) = \begin{bmatrix} 1-t & t-1 \\ 1-t & 0 \end{bmatrix}, \quad \aleph_{\sigma_1^2\sigma_2\sigma_1} = \begin{bmatrix} \frac{\sqrt{2}}{2} \\ \frac{\sqrt{2}}{2}(1-t) \end{bmatrix}, \tag{4.28}$$

$$\mathcal{B}(\sigma_1^2\sigma_2^2\sigma_1) = \begin{bmatrix} 0 & t \\ -t & t\frac{\sqrt{2}}{2}t \end{bmatrix}, \quad \aleph_{\sigma_1^2\sigma_2^2\sigma_1} = \begin{bmatrix} \frac{\sqrt{2}}{2} \\ \frac{\sqrt{2}}{2}(1-t) \end{bmatrix}, \tag{4.29}$$

$$\mathcal{B}(\sigma_2^2\sigma_1^2\sigma_2) = \begin{bmatrix} t & -1 \\ t-1 & 0 \end{bmatrix}, \quad \aleph_{\sigma_2^2\sigma_1^2\sigma_2} = \begin{bmatrix} \frac{\sqrt{2}}{2} \\ \frac{\sqrt{2}}{2}(t-1) \end{bmatrix}. \tag{4.30}$$

Only the subspaces, which are invariant under an action of the Burau representation on the complex space $\mathcal{C}^2$, are complex lines

$$z_1 = (t+1)z_1, \tag{4.31}$$

$$z_2 = \tfrac{1}{3}(t+1)z_1, \tag{4.32}$$

$$z_3 = (1-t)z_1, \tag{4.33}$$

$$z_4 = (t-1)z_1. \tag{4.34}$$

The isotropy groups[3] of these complex lines are the subgroups $\mathcal{Z}_3$ of the Burau representation $\mathcal{B}(B_3/\Omega_3^3)$ which are generated by elements: $e\sigma_1 e^{-1} = \sigma_1$, $(\sigma_1\sigma_2\sigma_1)\sigma_1(\sigma_1\sigma_2\sigma_1)^{-1} = \sigma_2$, $(\sigma_1^2\sigma_2)\sigma_1(\sigma_1^2\sigma_2)^{-1} = \sigma_1\sigma_2\sigma_1^2$, $(\sigma_1\sigma_2^2)\sigma_1(\sigma_1\sigma_2^2)^{-1} = \sigma_1^2\sigma_2\sigma_1$, respectively. The element σ_1 and the generators of the isotropy groups in the cases (4.31), (4.32), and (4.33) are conjugate by the generators k, j, i of the quaternion group $\mathcal{Q} \subset B_3/\Omega_3^3$. The complex lines (4.31)–(4.34) contain all fixed

[3]The isotropy group of the point $p \in M$ is a subgroup of the group G acting on the space M defined by

$$H(p) = \{g \in G : gp = p\}.$$

points of the representation $\mathcal{B}(B_3/\Omega_3^3)$ acting on the space $\mathcal{C}^2$ and, therefore, the representation B_3/Ω_3^3 acts freely[4] on the complement of these four complex lines.

The generators of the representation $\mathcal{B}(\sigma_1)$ i $\mathcal{B}(\sigma_2)$ act in the complex vector space $\mathcal{C}^2$ and they are of a finite order. They have eigenvalues 1 and $-t = \sqrt[3]{1}$, and then they are complex reflections (Hiller 1986). Let us note that the factor group B_3/Ω_3^3 has a faithful complex representation $\mathcal{B}(B_3/\Omega_3^3)$ and is generated by complex reflections while acting on a complex vector space $\mathcal{C}^2$. Therefore, B_3/Ω_3^3 is a complex reflection group (Hiller 1986). For such groups:

$$|B_3/\Omega_3^3| = \prod_{i=1}^{n} d_i, \tag{4.35}$$

$$M = \sum_{i=1}^{n}(d_i - 1), \tag{4.36}$$

where d_i are degrees of the invariant polynomials B_3/Ω_3^3, M is a number of complex reflections in $\mathcal{B}(B_3/\Omega_3^3)$, n is a number of invariant polynomials of the group B_3/Ω_3^3 (Springer 1977). According to the Chevalley theorem (Hiller 1986) for a complex reflection group, the number of invariant polynomials is equal to the dimension of the representation, and then there are two invariant polynomials of group B_3/Ω_3^3. The second eigenvalues of the matrices (4.23)–(4.30) are equal to $-t$ or $t-1$, that is, the third roots of 1, then all matrices (4.23)–(4.30) are complex reflections. The relations (4.35) and (4.36) take the form

$$d_1 \cdot d_2 = 24, \tag{4.37}$$

$$d_1 + d_2 = 10, \tag{4.38}$$

and we have $d_1 = 4$ and $d_2 = 6$. Let us denote the homogeneous polynomial of two variables over a ring of complex numbers $\mathcal{C}$, with degrees 4 and 6, by $\mathcal{F}_4(z_1, z_2)$ and $\mathcal{F}_6(z_1, z_2)$ in the form:

$$\mathcal{F}_4(z_1, z_2) = \sum_{i=0}^{4} a_i z_1^{4-i} z_2^i, \tag{4.39}$$

$$\mathcal{F}_6(z_1, z_2) = \sum_{i=0}^{6} b_i z_1^{6-i} z_2^i. \tag{4.40}$$

Let f be a map

$$f : (z_1, z_2) \to (Z_1, Z_2) = (\mathcal{F}_4, \mathcal{F}_6). \tag{4.41}$$

The image of the complex lines (4.31)–(4.32) is the set given by the parametric equations: $Z_1 = \alpha_1 z_1^4$, $Z_2 = \alpha_2 z_2^6$, $\alpha_1, \alpha_2 \in \mathcal{C}$, that is, the set of points satisfying the equation $Z_1^3 = \alpha Z_2^2$, $\alpha \in \mathcal{C}$. An intersection of this set with

[4]That is, with no fixed points.

a three-dimensional sphere S^3 is a $\mathcal{T}$ torus knot of type $(2,3)$ called a trefl (Milnor 1968). The complex lines (4.31)–(4.34) are hyperplanes in R^4 and then their intersections with the sphere S^3 are four circles S^1. The factor group B_3/Ω_3^3 acts freely in the complement of four complex lines (4.31)–(4.34), hence,

$$\begin{aligned}\mathcal{C}^2/(B_3/\Omega_3^3) &= R_+ \times \{[S^3 \setminus 4S^1]/(B_3/\Omega_3^3)\}\\ &= R_+ \times \{S^3 \setminus [4S^1/(B_3/\Omega_3^3)]\}.\end{aligned}$$

Finally,

$$\begin{aligned}&R^2 \times R_+ \times [(S^3 \setminus 4S^1)/(B_3/\Omega_3^3)]\\ &\quad = R^2 \times R_+ \times [S^3 \setminus \mathcal{T}] = Q_3(R^2),\end{aligned} \tag{4.42}$$

that is, the space

$$\Gamma_3^3(R^2) = R^2 \times R_+ \times [S^3 \setminus 4S^1] \tag{4.43}$$

is a regular covering space of the configuration space $Q_3(R^2)$.

For a system of three looped particles of the 5th order, the factor group B_3/Ω_3^5 has 600 elements and the fibre of the covering projection has 600 points; for $n > 5$, the factor group B_n/Ω_3^n is the group of an infinite order.

5

MANY-BODY METHODS FOR CHERN–SIMONS SYSTEMS

5.1 Random phase approximation for an anyon gas

In this chapter, the quantum theory of many-body Chern–Simons systems will be considered. As it was mentioned in Chapter 1, the Chern–Simons field is useful in formal description of the change of statistics in two-dimensional systems, especially when systems of anyons or composite fermions are considered. The system of anyons is recognized by means of the phase change of the many-particle wave function when two particles positions are exchanged:

$$\Psi \longrightarrow e^{i\theta}\Psi'. \tag{5.1}$$

The phase change $\theta = 0$ refers to bosons and $\theta = \pi$ refers to fermions. Composite fermions are described by the phase factor $\theta = (2p+1)\pi$ for p being an integer. For free particles, the Hamiltonian of the system contains only the kinetic term (spins are omitted)[1]

$$H_{\mathrm{s}} = \frac{1}{2m}\sum_{j=1}^{N} \mathbf{p}_j^2. \tag{5.2}$$

The many-particle wave function satisfies the following Schrödinger equation:

$$H_{\mathrm{s}}\Psi_{\mathrm{s}} = E\Psi_{\mathrm{s}}. \tag{5.3}$$

Although the Hamiltonian is of a simple form, it is not easy to solve this equation due to the complexity of the property (5.1). The many-particle wave function property (5.1) can be formally expressed (using the representation of particle position in complex number notation $z_j = x_j + iy_j$) as (Hanna *et al.* 1989):

$$\Psi_{\mathrm{s}}(z_1, z_2, \ldots, z_N) = \prod_{j<k} \frac{(z_j - z_k)^{(\theta-\pi)/\pi}}{|z_j - z_k|^{(\pi-\theta)/\pi}} \Psi_{\mathrm{F}}(z_1, z_2, \ldots, z_N), \tag{5.4}$$

where Ψ_{F} is the fermion function (antisymmetric). One should note that the complex term on the right-hand side of Eqn (5.4) marks only how the wave

[1]In real-life systems, this refers to a total spin polarization of the system. In $2+1$ time–space, spin is determined by representations of the rotation group $SO(2)$ which is isomorphic with U(1). In this case, all rotations commutate and the spin is not quantized. As, on the plane, irreducible unitary one-dimensional representations of the braid group coincide with the rotation group representations, the relationship between the spin and statistics is $s = \hbar\theta/2\pi$, s is spin, θ is the statistics parameter (Balachandran *et al.* 1990).

function phase changes when particles are transposed. It is also possible to split the function into a boson function and an appropriately modified complex term. In further considerations, the fermion representation will be used. This representation seems to be more convenient for anyon systems because for $\theta \neq 0$, a statistical (related to quantum statistics) interaction always exists (by analogy to exchange interaction). The fermion representation, as we shall see, brings many problems when a system of bosons needs to be given in terms of it, as it is significantly different from the initial one (in this case perturbation methods fail; Hanna *et al.* (1989)). Introducing the wave function (5.4) to the Schrödinger equation, the gauge transformation (related to the idea of Aharonov–Bohm effect) can be performed to express the equivalent equation in the fermion representation (with unchanged eigenvalues)

$$H\Psi_{\mathrm{F}} = E\Psi_{\mathrm{F}}. \tag{5.5}$$

In the notation of the second quantization representation and, additionally, after introduction of the external magnetic field, the Hamiltonian for free anyons system in the fermion representation is (Chen *et al.* 1989; Hanna *et al.* 1989):

$$H = \int d^2\mathbf{r}\Psi^+(\mathbf{r})\frac{1}{2m}\left(\mathbf{p} + \frac{e}{c}\mathbf{A}(\mathbf{r}) + \frac{e}{c}\mathbf{A}^{\mathrm{ex}}(\mathbf{r})\right)^2\Psi(\mathbf{r}), \tag{5.6}$$

where Ψ is the fermion (spinless) field operator, $\mathbf{A}^{\mathrm{ex}}$ is the vector potential of the external magnetic field. The gauge field $\mathbf{A}$, called the Chern–Simons field,[2] is described as:

$$A_\alpha(\mathbf{r}) = \frac{(\pi - \theta)}{\pi}\frac{\hbar c}{e}\int d^2\mathbf{r}\epsilon_{\alpha\beta}\frac{(\mathbf{r} - \mathbf{r}')_\beta}{|\mathbf{r} - \mathbf{r}'|^2}\rho(\mathbf{r}), \tag{5.7}$$

where $\rho(\mathbf{r}) = \Psi^+(\mathbf{r})\Psi(\mathbf{r})$.

As zeroth order approximation (unperturbed system), the mean field Hamiltonian is considered:

$$H_0 = \int d^2\mathbf{r}\Psi^+(\mathbf{r})\frac{1}{2m}\left(\mathbf{p} + \frac{e}{c}\bar{\mathbf{A}}(\mathbf{r}) + \frac{e}{c}\mathbf{A}^{ex}(\mathbf{r})\right)^2\Psi(\mathbf{r}). \tag{5.8}$$

This approximation consists in averaging of point magnetic fluxes (introduced by the Chern–Simons field) in such a way that the total flux coming through the given surface does not change. Thus, the fictitious uniform statistical field (of magnetic type) $B^{\mathrm{s}} = \Phi^{\mathrm{s}}/S = (\pi - \theta)\bar{\rho}(hc/e)$ is introduced. In the Hamiltonian H_0, the mean field is determined by the vector potential defined by the following integral equation:

$$\bar{A}_\alpha(\mathbf{r}) = \frac{(\pi - \theta)}{\pi}\frac{\hbar c}{e}\int d^2\mathbf{r}\epsilon_{\alpha\beta}\frac{(\mathbf{r} - \mathbf{r}')_\beta}{|\mathbf{r} - \mathbf{r}'|^2}\bar{\rho}, \tag{5.9}$$

[2]Electrodynamics in two-dimension allows for reducing the number of Maxwell equations to three, but an ambiguity of solutions appears which can be removed by introducing the Chern–Simons term (Jackiv 1989).

where $\bar{\rho}$ is the average particle density of the system. The integral (5.9) is divergent, however, if calculated over a finite area gives the expected result $B^s = \nabla \times \bar{\mathbf{A}}$ (Chen *et al.* 1989; Jacak and Sitko 1993). Due to the Hamiltonian form for free anyons (5.2), the commutation relation $[p_x, p_y] = 0$ is satisfied. However, the introduction of the mean field breaks this relation: $[p_x, p_y] \neq 0$ (but still $[p_x, H] = [p_y, H] = 0$ (Wilczek 1990), similarly as in the case of systems in an external magnetic field). This situation is often compared to a spontaneous symmetry breaking (in this case it is a spontaneous breaking of the commutation rule; Chen *et al.* 1989).

The remaining part of the Hamiltonian (5.6)

$$H_{\rm int} = H - H_0 \tag{5.10}$$

is treated as the perturbation which can be interpreted as the interaction although the system of free particles is considered. The interaction Hamiltonian, besides a two-particle term, contains a non-standard three-particle term as well. The presence of the three-particle interaction makes the model more complicated. It is easy to note that such an interaction entails the necessity of introducing a rich diagram analysis and approximations must be used. The most widely used and the most productive is the random phase approximation (RPA) (Chen *et al.* 1989; Fetter *et al.* 1989), which will be discussed further.

The interaction Hamiltonian can be expressed in the following form:

$$\begin{aligned} H_{\rm int} &= H - H_0 \\ &= \frac{e}{2mc} \int d^2\mathbf{r}\Psi^+(\mathbf{r}) \left[2\left(\mathbf{p} + \frac{e}{c}\bar{\mathbf{A}} + \frac{e}{c}\mathbf{A}^{\rm ex}\right)(\mathbf{A} - \bar{\mathbf{A}}) + \frac{e}{c}(\mathbf{A} - \bar{\mathbf{A}})^2 \right] \Psi(\mathbf{r}) \\ &= H_1 + H_2, \end{aligned} \tag{5.11}$$

where

$$\begin{aligned} H_1 &= \frac{(\pi - \theta)}{\pi} \frac{\hbar}{m} \iint d^2\mathbf{r} d^2\mathbf{r}' \Psi^+(\mathbf{r}) \left(p_\alpha + \frac{e}{c}\bar{A}_\alpha + \frac{e}{c}A_\alpha^{\rm ex} \right) \Psi(\mathbf{r}) \\ &\quad \times \epsilon_{\alpha\beta} \frac{(\mathbf{r} - \mathbf{r}')_\beta}{|\mathbf{r} - \mathbf{r}'|^2} (\rho(\mathbf{r}') - \bar{\rho}), \end{aligned} \tag{5.12}$$

$$\begin{aligned} H_2 &= \frac{(\pi - \theta)^2}{\pi^2} \frac{\hbar^2}{2m} \iiint d^2\mathbf{r} d^2\mathbf{r}' d^2\mathbf{r}'' \rho(\mathbf{r}) \frac{(\mathbf{r} - \mathbf{r}')}{|\mathbf{r} - \mathbf{r}'|^2} \frac{(\mathbf{r} - \mathbf{r}'')}{|\mathbf{r} - \mathbf{r}''|^2} \\ &\quad \times (\rho(\mathbf{r}') - \bar{\rho})(\rho(\mathbf{r}'') - \bar{\rho}). \end{aligned} \tag{5.13}$$

The term H_2 includes the three-particle interaction. The effective field operator is defined as follows (effective magnetic field: $B^* = B^{\mathrm{ex}} + B^{\mathrm{s}}$)

$$\hat{\jmath}^{\alpha} = \Psi^{+}(\mathbf{r})\frac{1}{m}\left(p_{\alpha} + \frac{e}{c}\bar{A}_{\alpha}(\mathbf{r}) + \frac{e}{c}A^{\mathrm{ex}}_{\alpha}(\mathbf{r})\right)\Psi(\mathbf{r}), \;\; \hat{\jmath}^{0} = \rho(\mathbf{r}) - \bar{\rho}.$$

If in the term H_2 one puts $\rho(\mathbf{r}) = \bar{\rho} + (\rho(\mathbf{r}) - \bar{\rho})$, the interaction Hamiltonian can be expressed as follows:

$$H_{\mathrm{int}} = \tfrac{1}{2}\int d^2\mathbf{r}\int d^2\mathbf{r}' j^{\mu}(\mathbf{r})V_{\mu\nu}(\mathbf{r},\mathbf{r}')j^{\nu}(\mathbf{r}') + H_3,$$

where H_3 denotes the three-particle term. Using the Fourier transformation,

$$\int d^2\mathbf{r}\frac{\mathbf{r}}{|\mathbf{r}|^2}e^{-i\mathbf{q}\cdot\mathbf{r}} = \frac{2\pi}{i}\frac{\mathbf{q}}{|\mathbf{q}|^2}, \tag{5.14}$$

the interaction matrix in the Fourier image can be expressed as follows $(\mu, \nu = 0, x, y)$

$$V(\mathbf{q}) = \frac{2(\pi-\theta)}{q^2}\begin{pmatrix} \dfrac{(\pi-\theta)}{\pi}\dfrac{h\bar{\rho}}{m} & iq_y & -iq_x \\ -iq_y & 0 & 0 \\ iq_x & 0 & 0 \end{pmatrix}.$$

Within the framework of the RPA in Chern–Simons systems, the three-particle term H_3 is omitted as proportional to the cube of the density fluctuation. This agrees with the idea of the RPA: (Pines and Nozieres 1966)

$$\rho_{\mathbf{k}-\mathbf{q}} = \sum_{j=1}^{N} e^{-i(\mathbf{k}-\mathbf{q})\cdot\mathbf{r}_j} \approx N\delta_{\mathbf{k},\mathbf{q}}. \tag{5.15}$$

Therefore, in the adopted approximation, the interaction reduces to the expression for H_1 and the logarithmic interaction. The presence of the logarithmic interaction (its Fourier transformation brings the term $1/q^2$) is a very important characteristic of the system. Moreover, it indicates possible analogies with the Kosterlitz–Thouless model of the phase transition in two-dimension, which will be discussed in more detail later. Such interaction formally resembles the Coulomb interaction in two-dimension and it leads to different effects typical of electric interaction (e.g. to a Meissner effect) although the real electric interaction of charged particles has not been taken into account yet.[3] Now, the linear response of Chern–Simons systems to the external electromagnetic perturbation a^{ex}_{ν} (potential of the 2 + 1 electromagnetic field, $\nu = 0, x, y$) will be considered.

[3] One should note that when considering the Chern–Simons theory one has charged particles and the field couples to the charge. Moreover, the Aharonov–Bohm effect also illustrates this fact, as the phase change can be expressed as a product of the charge and flux attached to the particle.

This response is expressed by means of the Kubo (Fetter and Walecka 1971) relation:

$$e\langle \hat{J}^{\mu}(\mathbf{q},\omega)\rangle = \frac{c}{4\pi} K^{\mu\nu}(\mathbf{q},\omega) a_{\nu}^{\mathrm{ex}}(\mathbf{q},\omega), \tag{5.16}$$

where

$$K^{\mu\nu}(\mathbf{q},\omega) = \frac{4\pi\bar{\rho}e^2}{mc^2}\delta^{\mu\nu}(1-\delta^{\mu 0}) + \frac{4\pi e^2}{\hbar c^2}\Delta_R^{\mu\nu}(\mathbf{q},\omega) \tag{5.17}$$

and Δ_R is the retarded correlation function of currents. In the real space representation, this function is

$$\Delta_R^{\mu\nu}(\mathbf{r}t,\mathbf{r}'t') = -i\theta(t-t')\langle[\hat{J}^{\mu}(\mathbf{r}t), \hat{J}^{\nu}(\mathbf{r}'t')]\rangle, \tag{5.18}$$

where the current operator is given by:

$$\hat{J}^{\alpha} = \Psi^{+}(\mathbf{r})\frac{1}{m}\left(p_{\alpha} + \frac{e}{c}A_{\alpha}(\mathbf{r}) + \frac{e}{c}A_{\alpha}^{\mathrm{ex}}(\mathbf{r})\right)\Psi(\mathbf{r}). \tag{5.19}$$

Eqn (5.16) also describes the system response to the scalar field a_0^{ex}, where $\hat{J}^0 = \rho - \bar{\rho} = \hat{\jmath}^0$. The so-called Coulomb gauge proves convenient in this case, $\nabla\cdot\mathbf{A} = 0$ (both the external and the mean Chern–Simons field undergo identical gauge), and $\mathbf{q} = q\hat{\mathbf{x}}$. Using the equation of continuity,

$$\mathbf{J}\cdot\mathbf{q} = \omega J^0, \tag{5.20}$$

one has $J^x = (\omega/q)J^0$, and one dimension can be eliminated when matrix $K^{\mu\nu}$ is calculated (μ, $\nu = 0$, y) (Halperin *et al.* 1993). Let the correlation function of effective currents be considered first:

$$D_R^{\mu\nu}(\mathbf{r}t,\mathbf{r}'t') = -i\theta(t-t')\langle[\hat{\jmath}^{\mu}(\mathbf{r}t), \hat{\jmath}^{\nu}(\mathbf{r}'t')]\rangle. \tag{5.21}$$

In RPA, only bubble diagrams[4] are considered (Fig. 5.1),

$$D_R^{\mathrm{RPA}}(\mathbf{q},\omega) = [I - D_R^0(\mathbf{q},\omega)V(\mathbf{q})]^{-1}D_R^0(\mathbf{q},\omega). \tag{5.22}$$

Now, the difference between the correlation function of the full currents Δ_R and the correlation function of the effective currents D_R will be considered. Additionally, in the Fourier image

$$\mathbf{J}_{\mathbf{q}} - \mathbf{j}_{\mathbf{q}} = L^{-2}\sum_{\mathbf{k}\neq 0}\mathbf{A}(\mathbf{k})\tilde{\rho}_{\mathbf{k}}\rho_{\mathbf{k}-\mathbf{q}}, \tag{5.23}$$

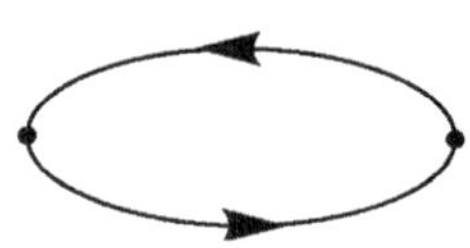

FIG. 5.1. The RPA bubble graph.

[4] D_R^0 is the density–density correlation function given by the bubble diagram—Fig. 5.1.

where L^2 is the area of the system and

$$\mathbf{A}(\mathbf{k}) = \frac{\pi - \theta}{\pi}\frac{\hbar c}{e}\frac{\hat{\mathbf{z}} \times \mathbf{q}}{|\mathbf{q}|^2}, \quad \tilde{\rho} = \rho - \bar{\rho}. \tag{5.24}$$

Within the RPA, only the contribution from the $\mathbf{k} = \mathbf{q}$ term is taken into account. Then,

$$\mathbf{J}_\mathbf{q} \simeq \mathbf{j}_\mathbf{q} + \bar{\rho}\mathbf{A}(\mathbf{q})\tilde{\rho}_\mathbf{q}. \tag{5.25}$$

The difference between the correlation functions can be expressed by the following equation (Fetter *et al.* 1991):

$$\Delta_R \simeq \Delta_R^{\mathrm{RPA}} = (I + U^+)D_R^{\mathrm{RPA}}(I + U), \tag{5.26}$$

where

$$U = \frac{(\pi - \theta)(h\bar{\rho}/\pi m)}{q}\begin{pmatrix} 0 & -i \\ 0 & 0 \end{pmatrix}. \tag{5.27}$$

Thus, in the case of Chern–Simons systems, the RPA consists both in taking of sum of bubble diagrams into account and in reducing the interaction Hamiltonian to the two-body terms. This approach is equivalent to the self-consistent Hartree approximation (Fetter and Hanna 1992*a*), which will be discussed in Section 5.4.

5.2 Correlation energy of an anyon gas

As the interaction can be expressed in the matrix form, it is interesting to find suitable correlation energy equations (which, next, will be used in energy calculations in Chapters 6 and 7). If E_0 is the ground-state energy in the zeroth order approximation (mean field approximation) and E_1 is the Hartree–Fock correction (the interaction average in the unperturbed state), then the ground-state energy of the system looks as follows:

$$E = E_0 + E_1 + E_{\mathrm{c}}, \tag{5.28}$$

where E_{c} is the correlation energy. When introducing the coupling parameter (for an electron gas it can be a charge of an electron; Pines and Nozieres (1966)), the following expression is found:

$$E_{\mathrm{c}} = \int_0^1 \frac{d\lambda}{\lambda}(\langle \lambda H_{\mathrm{int}}\rangle_\lambda - \langle \lambda H_{\mathrm{int}}\rangle_0). \tag{5.29}$$

We are going to express the interaction Hamiltonian (5.11) in a different form (symmetric with respect to integral variables). Using the commutation relations for fermion field operators, one expresses the interaction Hamiltonian by means of density and current operators:

$$
\begin{aligned}
H_1 = & \int d\mathbf{r}_1 \left(-\frac{e}{c}\bar{\mathbf{A}}_1 \cdot \mathbf{j}_1 + \frac{e^2}{2c^2}|\bar{\mathbf{A}}_1|^2 \rho_1 \right) \\
& + \frac{e}{2c} \int d\mathbf{r}_1 \int d\mathbf{r}_2 [\mathbf{A}_{12} \cdot \mathbf{j}_1 \rho_2 + \rho_1 \mathbf{j}_2 \cdot \mathbf{A}_{12} - \delta(\mathbf{r}_1 - \mathbf{r}_2)(\mathbf{A}_{12} \cdot \mathbf{j}_1 + \mathbf{A}_{21} \cdot \mathbf{j}_2)] \\
& + \frac{e^2}{mc^2} \int d\mathbf{r}_1 \int d\mathbf{r}_2 (|\mathbf{A}_{12}|^2 - \mathbf{A}_{12} \cdot \bar{\mathbf{A}}_1 - \mathbf{A}_{21} \cdot \bar{\mathbf{A}}_2)[\rho_1 \rho_2 - \delta(\mathbf{r}_1 - \mathbf{r}_2)\rho_1],
\end{aligned} \tag{5.30}
$$

$$
\begin{aligned}
H_2 = & \tfrac{1}{6} \iiint d\mathbf{r}_1 d\mathbf{r}_2 d\mathbf{r}_3 \Psi^+(\mathbf{r}_1)\Psi^+(\mathbf{r}_2)\Psi^+(\mathbf{r}_3) \\
& \times w(\mathbf{r}_1, \mathbf{r}_2, \mathbf{r}_3)\Psi(\mathbf{r}_3)\Psi(\mathbf{r}_2)\Psi(\mathbf{r}_1),
\end{aligned} \tag{5.31}
$$

$\rho_1 = \rho(\mathbf{r}_1)$. The integral kernel w in the term H_2 (a symmetric form of the three-particle interaction) is:

$$
w(\mathbf{r}_1, \mathbf{r}_2, \mathbf{r}_3) = \frac{e^2}{mc^2}(\mathbf{A}_{12} \cdot \mathbf{A}_{13} + \mathbf{A}_{21} \cdot \mathbf{A}_{23} + \mathbf{A}_{31} \cdot \mathbf{A}_{32}). \tag{5.32}
$$

The average of the interaction Hamiltonian is given by (Hanna and Fetter 1993):

$$
\begin{aligned}
\langle H_{\text{int}} \rangle = & \bar{\rho}\frac{e^2}{mc^2} \int d\mathbf{r}_1 \int d\mathbf{r}_2 \delta(\mathbf{r}_1 - \mathbf{r}_2)(-|\mathbf{A}_{12}|^2 + \mathbf{A}_{12} \cdot \bar{\mathbf{A}}_1 + \mathbf{A}_{21} \cdot \bar{\mathbf{A}}_2) \\
& + \frac{1}{3}\bar{\rho} \iiint d\mathbf{r}_1 d\mathbf{r}_2 d\mathbf{r}_3 \delta(\mathbf{r}_1 - \mathbf{r}_2)\delta(\mathbf{r}_1 - \mathbf{r}_3) w(\mathbf{r}_1, \mathbf{r}_2, \mathbf{r}_3) \\
& + \frac{1}{2} \iint d\mathbf{r}_1 d\mathbf{r}_2 \left[\frac{e}{c}\mathbf{A}_{12} \cdot \langle \mathbf{j}_1 \tilde{\rho}_2 \rangle + \frac{e}{c}\langle \tilde{\rho}_1 \mathbf{j}_2 \rangle \cdot \mathbf{A}_{12} + w_0(\mathbf{r}_1, \mathbf{r}_2)\langle \tilde{\rho}_1 \tilde{\rho}_2 \rangle \right] \\
& + \frac{1}{6} \iiint d\mathbf{r}_1 d\mathbf{r}_2 d\mathbf{r}_3 w(\mathbf{r}_1, \mathbf{r}_2, \mathbf{r}_3)\langle \tilde{\rho}_1 \tilde{\rho}_2 \tilde{\rho}_3 \rangle,
\end{aligned} \tag{5.33}
$$

where $\tilde{\rho}_1 = \rho(\mathbf{r}_1) - \bar{\rho}$, and

$$
w_0(\mathbf{r}_1, \mathbf{r}_2) = \frac{e^2}{mc^2}\bar{\rho} \int d\mathbf{r}_3 \mathbf{A}_{31} \cdot \mathbf{A}_{32} \tag{5.34}
$$

is the logarithmic interaction, extracted from the H_2 term. After subtracting the interaction average in the unperturbed state from the Eqn (5.33), the following expression is obtained (the first two terms in Eqn (5.33) disappear):

$$\begin{aligned}\langle H_{\text{int}}\rangle - \langle H_{\text{int}}\rangle_0 = \frac{1}{2}\iint d\mathbf{r}_1 d\mathbf{r}_2 \Big[\frac{e}{c}\mathbf{A}_{12}\cdot(\langle \mathbf{j}_1\tilde{\rho}_2\rangle - \langle \mathbf{j}_1\tilde{\rho}_2\rangle_0)\\ +\frac{e}{c}(\langle\tilde{\rho}_1\mathbf{j}_2\rangle - \langle\tilde{\rho}_1\mathbf{j}_2\rangle_0)\cdot\mathbf{A}_{12} + w_0(\mathbf{r}_1,\mathbf{r}_2)(\langle\tilde{\rho}_1\tilde{\rho}_2\rangle - \langle\tilde{\rho}_1\tilde{\rho}_2\rangle_0)\Big]\\ +\frac{1}{6}\iiint d\mathbf{r}_1 d\mathbf{r}_2 d\mathbf{r}_3 w(\mathbf{r}_1,\mathbf{r}_2,\mathbf{r}_3)(\langle\tilde{\rho}_1\tilde{\rho}_2\tilde{\rho}_3\rangle - \langle\tilde{\rho}_1\tilde{\rho}_2\tilde{\rho}_3\rangle_0).\end{aligned} \tag{5.35}$$

Unlike an electron gas, for which the correlation energy can be expressed only by means of the density–density correlation function, for anyon gas the contributions from both the current–density correlation function and the density–density–density three-particle correlation function must be taken into account. The time-ordered correlation function is defined as follows:

$$\begin{aligned}\hbar D^{\mu\nu}(\mathbf{r}_1 t_1, \mathbf{r}_2 t_2) &= -i\langle T[j^\mu(\mathbf{r}_1 t_1) j^\nu(\mathbf{r}_2 t_2)]\rangle\\ &= \hbar L^{-2}\sum_{\mathbf{q}\neq 0}\int_{-\infty}^{\infty}\frac{d\omega}{2\pi}e^{i(\mathbf{q}\cdot\mathbf{r}_{12}-\omega t_{12})}D^{\mu\nu}(\mathbf{q},\omega).\end{aligned} \tag{5.36}$$

In the expression (5.35) one has the averages of products of density and current operators. To obtain them from the time-ordered function (5.36), the imaginary term $D^{\mu\nu}$ will be considered:

$$\hbar\int_0^\infty \frac{d\omega}{\pi}\text{Im}D^{\mu\nu}(\mathbf{q},\omega) = -L^2\langle 0|j^\mu_{\mathbf{q}} j^\nu_{-\mathbf{q}}|0\rangle. \tag{5.37}$$

Thus, using the matrix form of the interaction one can write

$$\begin{aligned}E_c = -\frac{1}{2}\hbar\frac{N}{\rho}\sum_{\mathbf{q}\neq 0}\int_0^\infty\frac{d\omega}{\pi}\int_0^1\frac{d\lambda}{\lambda}\text{Im Tr}\left\{\lambda V(q)[D_\lambda^{RPA}(\mathbf{q},\omega) - D_0(\mathbf{q},\omega)]\right\}\\ +\frac{1}{6}\iiint d\mathbf{r}_1 d\mathbf{r}_2 d\mathbf{r}_3\int_0^1\frac{d\lambda}{\lambda}\lambda w(\mathbf{r}_1,\mathbf{r}_2,\mathbf{r}_3)(\langle\tilde{\rho}_1\tilde{\rho}_2\tilde{\rho}_3\rangle_\lambda - \langle\tilde{\rho}_1\tilde{\rho}_2\tilde{\rho}_3\rangle_0).\end{aligned} \tag{5.38}$$

After omitting the three-particle term and determining correlation function within the RPA:

$$D_\lambda^{\text{RPA}}(\mathbf{q},\omega) = [I - \lambda D_0(\mathbf{q},\omega)V(\mathbf{q})]^{-1}D_0(\mathbf{q},\omega), \tag{5.39}$$

the following equation is obtained:

$$\begin{aligned}E_c^{\text{RPA}} = -\frac{1}{2}\hbar\frac{N}{\rho}\int\frac{d\mathbf{q}}{(2\pi)^2}\int_0^\infty\frac{d\omega}{\pi}\int_0^1\frac{d\lambda}{\lambda}\\ \times\text{Im Tr}\left\{\lambda V(q)[D_\lambda^{\text{RPA}}(\mathbf{q},\omega) - D_0(\mathbf{q},\omega)]\right\}.\end{aligned} \tag{5.40}$$

The above expression will be used while considering the fractional quantum Hall effect (FQHE) to determine the ground state energy of composite fermion system.

5.3 Hartree–Fock approximation for Chern–Simons systems

In the Hartree–Fock approximation, which we consider now, the description of a many-particle system is reduced to the effective description of a single-particle system, that is, it is assumed that a particle moves in the averaged interaction potential of other particles. Moreover, each particle contributes to the interaction in a self-consistent way. This results in the following system of the Hartree–Fock equations:

$$H_{\mathrm{HF}}\phi_i = E_i\phi_i, \tag{5.41}$$

where H_{HF} is the effective Hamiltonian of a single particle in the Hartree–Fock approximation, ϕ_i are single-particle eigenstates. Generally, the Hartree–Fock method consists of self-consistent determination of single-particle wave functions, energy eigenvalues, and the single-particle Hamiltonian H_{HF}. However, in the case of an electron gas, the self-consistency is satisfied by plane waves, and the Hartree–Fock approximation is equivalent to the first order perturbation method (Pines and Nozieres 1966; Fetter and Walecka 1971). It will also be shown that in the case of the homogeneous anyon gas, the single-particle Hartree–Fock states are Landau levels. According to the idea of the Hartree–Fock approximation, the system wave function is assumed to be the Slater determinant of single-particle states:

$$\Phi(\mathbf{r}_1, \mathbf{r}_2, ..., \mathbf{r}_N) = \frac{1}{\sqrt{N!}}\sum_{\sigma}^{N!} \mathrm{sgn}(\sigma)\phi_{\sigma(1)}(\mathbf{r}_1)\phi_{\sigma(2)}(\mathbf{r}_2)\cdots\phi_{\sigma(N)}(\mathbf{r}_N), \tag{5.42}$$

where the summation goes over all possible permutations of N elements. Let the Hamiltonian of the Chern–Simons system be expressed as the sum of the following terms:

$$H_1 = \frac{1}{2m}\sum_l \left(\mathbf{p}_l + \frac{e}{c}\mathbf{A}_l^{\mathrm{ex}}\right)^2, \tag{5.43}$$

$$H_{2a} = \frac{e}{mc}\sum_{l,m\neq l} \left(\mathbf{p}_l + \frac{e}{c}\mathbf{A}_l^{\mathrm{ex}}\right)\mathbf{A}_{lm}, \tag{5.44}$$

$$H_{2b} = \frac{e^2}{2mc^2}\sum_{l,m\neq l} (\mathbf{A}_{lm})^2, \tag{5.45}$$

$$H_3 = \frac{e^2}{2mc^2}\sum_{l,m\neq l,k\neq l,m} \mathbf{A}_{lm}\mathbf{A}_{lk}, \tag{5.46}$$

where $\mathbf{A}_l = \sum_{m\neq l}\mathbf{A}_{lm}$. The average of the Hamiltonian $\langle\Phi|H|\Phi\rangle$, assuming that single-particle states are orthonormal, is:

$$\langle H_1\rangle = \frac{1}{2m}\sum_l \int d1\phi_l^+(1)\left(\mathbf{p}_1 + \frac{e}{c}\mathbf{A}_1^{\mathrm{ex}}\right)^2\phi_l(1), \tag{5.47}$$

$$\langle H_{2a}\rangle = \frac{e}{mc}\sum_{l,m\neq l}\int d1\, d2\, \phi_l^+(1)\phi_m^+(2)\left(\mathbf{p}_1 + \frac{e}{c}\mathbf{A}_1^{\mathrm{ex}}\right)\mathbf{A}_{12}$$
$$\times\,[\phi_l(1)\phi_m(2) - \phi_l(2)\phi_m(1)], \tag{5.48}$$

$$\langle H_{2b}\rangle = \frac{e^2}{2mc^2}\sum_{l,m\neq l}\int d1\, d2\, \phi_l^+(1)\phi_m^+(2)(\mathbf{A}_{12})^2$$
$$\times\,[\phi_l(1)\phi_m(2) - \phi_l(2)\phi_m(1)], \tag{5.49}$$

where $d1 = d^2\mathbf{r}_1$, indexes l, m number quantum states, summation goes over occupied states. Two terms, typical for the Hartree–Fock approximation, appearing in the Eqs (5.48) and (5.49) are the Hartree term and exchange (Fock) term. The analogous expression for the three-particle interaction contains 3! of different contributions:

$$\langle H_3\rangle = \frac{e^2}{2mc^2}\sum_{l,m\neq l,k\neq l,m}\int d1\, d2\, d3\, \phi_l^+(1)\phi_m^+(2)\phi_k^+(3)\mathbf{A}_{lm}\mathbf{A}_{lk}$$
$$\times\left[\sum_{\sigma}^{3!}\mathrm{sgn}(\sigma)\phi_l(\sigma(1))\phi_m(\sigma(2))\phi_k(\sigma(3))\right]. \tag{5.50}$$

After simple transformations of equations, using the definition of the vector potential of the mean field (5.9), the average value of the Hamiltonian reduces to the following six expressions:

$$\langle H_a\rangle = \frac{1}{2m}\sum_l\int d1\, \phi_l^+(1)\left(\mathbf{P}_1 + \frac{e}{c}\bar{\mathbf{A}}_1 + \frac{e}{c}\mathbf{A}_1^{\mathrm{ex}}\right)^2\phi_l(1), \tag{5.51}$$

$$\langle H_b\rangle = -\frac{e}{mc}\sum_{l,m}\int d1\, d2\phi_l^+(1)\phi_m^+(2)\mathbf{A}_{12}\left(\mathbf{P}_1 + \frac{e}{c}\bar{\mathbf{A}}_1 + \frac{e}{c}\mathbf{A}_1^{\mathrm{ex}}\right)\phi_l(2)\phi_m(1), \tag{5.52}$$

$$\langle H_c\rangle = \frac{e^2}{2c^2m}\sum_{l,m}\int d1\, d2\, \phi_l^+(1)\phi_m^+(2)|\mathbf{A}_{12}|^2\phi_l(1)\phi_m(2), \tag{5.53}$$

$$\langle H_d\rangle = -\frac{e^2}{2c^2m}\sum_{l,m}\int d1\, d2\, \phi_l^+(1)\phi_m^+(2)|\mathbf{A}_{12}|^2\phi_l(2)\phi_m(1), \tag{5.54}$$

$$\langle H_e\rangle = \frac{e^2}{c^2m}\sum_{l,m,k}\int d1\, d2\, d3\, \phi_l^+(1)\phi_m^+(2)\phi_k^+(3)\mathbf{A}_{12}\mathbf{A}_{13}\phi_l(3)\phi_m(1)\phi_k(2), \tag{5.55}$$

$$\langle H_f\rangle = -\frac{e^2}{2c^2m}\sum_{l,m,k}\int d1\, d2\, d3\, |\phi_l(1)|^2\phi_m^+(2)\phi_k^+(3)\mathbf{A}_{12}\mathbf{A}_{13}\phi_m(3)\phi_k(2). \tag{5.56}$$

The Hamiltonian of a single particle, $H_{\rm HF}$, in the Hartree–Fock approximation is the first variation of the average energy $\langle H\rangle = \sum_{i=a}^{f}\langle H_i\rangle$ with respect to $\phi_l^+(1)$ (Hanna *et al.* 1989). The variation of the term $\langle H_a\rangle$ leads to equations:

$$H_{\rm HF}^1\phi_l(1) = \frac{1}{2m}\sum_m \int d2\phi_m^+(2)\left(\mathbf{P}_1 + \frac{e}{c}\bar{\mathbf{A}}_1 + \frac{e}{c}\mathbf{A}_1^{\rm ex}\right)^2 \phi_m(2)\phi_l(1), \qquad (5.57)$$

$$H_{\rm HF}^2\phi_l(1) = \frac{e}{mc}\sum_m \int d2\phi_m^+(2)\mathbf{A}_{21}\left(\mathbf{P}_2 + \frac{e}{c}\bar{\mathbf{A}}_2 + \frac{e}{c}\mathbf{A}_2^{\rm ex}\right)\phi_m(2)\phi_l(1). \qquad (5.58)$$

The variation of the term $\langle H_b\rangle$ results in:

$$H_{\rm HF}^3\phi_l(1) = -\frac{e}{mc}\sum_m \int d2\phi_m^+(2)\mathbf{A}_{12}\left(\mathbf{P}_1 + \frac{e}{c}\bar{\mathbf{A}}_1 + \frac{e}{c}\mathbf{A}_1^{\rm ex}\right)\phi_m(1)\phi_l(2), \qquad (5.59)$$

$$H_{\rm HF}^4\phi_l(1) = -\frac{e}{mc}\sum_m \int d2\phi_m^+(2)\mathbf{A}_{21}\left(\mathbf{P}_2 + \frac{e}{c}\bar{\mathbf{A}}_2 + \frac{e}{c}\mathbf{A}_2^{\rm ex}\right)\phi_m(1)\phi_l(2), \qquad (5.60)$$

$$H_{\rm HF}^5\phi_l(1) = -\frac{e^2}{mc^2}\sum_{m,k} \int d2d3\phi_m^+(2)\phi_k^+(3)\mathbf{A}_{32}\mathbf{A}_{31}\phi_m(3)\phi_k(2)\phi_l(1). \qquad (5.61)$$

Then,

$$H_{\rm HF}^6\phi_l(1) = \frac{e^2}{mc^2}\sum_m \int d2|\phi_m(2)|^2|\mathbf{A}_{12}|^2\phi_l(1), \qquad (5.62)$$

and

$$H_{\rm HF}^7\phi_l(1) = -\frac{e^2}{mc^2}\sum_m \int d2\phi_m^+(2)|\mathbf{A}_{12}|^2\phi_m(1)\phi_l(2) \qquad (5.63)$$

are found from variations of the terms $\langle H_c\rangle$ and $\langle H_d\rangle$. The first variation of the term $\langle H_e\rangle$ gives:

$$H_{\rm HF}^8\phi_l(1) = \frac{e^2}{mc^2}\sum_{m,k} \int d2\,d3\phi_m^+(3)\phi_k^+(2)\mathbf{A}_{12}\mathbf{A}_{13}\phi_m(1)\phi_k(3)\phi_l(2), \qquad (5.64)$$

$$H_{\rm HF}^9\phi_l(1) = \frac{e^2}{mc^2}\sum_{m,k} \int d2\,d3\phi_m^+(3)\phi_k^+(2)\mathbf{A}_{21}\mathbf{A}_{23}\phi_m(1)\phi_k(3)\phi_l(2), \qquad (5.65)$$

$$H_{\rm HF}^{10}\phi_l(1) = \frac{e^2}{mc^2}\sum_{m,k} \int d2\,d3\phi_m^+(3)\phi_k^+(2)\mathbf{A}_{31}\mathbf{A}_{32}\phi_m(1)\phi_k(3)\phi_l(2). \qquad (5.66)$$

The variation of the term $\langle H_f\rangle$ results in:

$$H_{\rm HF}^{11}\phi_l(1) = -\frac{e^2}{2mc^2}\sum_{m,k} \int d2\,d3\phi_m^+(2)\phi_k^+(3)\mathbf{A}_{12}\mathbf{A}_{13}\phi_m(3)\phi_k(2)\phi_l(1), \qquad (5.67)$$

$$H_{\rm HF}^{12}\phi_l(1) = -\frac{e^2}{mc^2}\sum_{m,k} \int d2\,d3\phi_m^+(2)|\phi_k(3)|^2\mathbf{A}_{31}\mathbf{A}_{32}\phi_m(1)\phi_l(2). \qquad (5.68)$$

After integrating over $\mathbf{r}_3$ in the last term, one can note that H^{12}_{HF} is the logarithmic interaction exchange term. The energy spectrum of single-particle excitations is obtained while determining eigenvalues of the Hamiltonian $H_{\mathrm{HF}} = \sum_{i=1}^{12} H^i_{\mathrm{HF}}$. This will be done in the following sections for different realizations of Chern–Simons systems.

5.4 Diagram analysis for a gas of anyons

5.4.1 *Self-consistent Hartree approximation for a gas of anyons*

In previous sections, the random phase and the Hartree–Fock approximations were presented in the form in which they will be used in the following chapters for anyons and composite fermions. It is also interesting to analyse those approximations in diagram representations. Using the split-up Hamiltonian of the Chern–Simons system expressed by the Eqns (5.43)–(5.46), let them be expressed in the second quantization formalism (the external magnetic field is omitted, for the sake of simplicity):

$$H_1 = \int d\mathbf{r}_1 \Psi^+(\mathbf{r}_1)\frac{\mathbf{p}_1^2}{2m}\Psi^+(\mathbf{r}_1), \tag{5.69}$$

$$H_{2a} = \frac{e}{2mc}\iint d\mathbf{r}_1 d\mathbf{r}_2 \Psi^+(\mathbf{r}_1)\Psi^+(\mathbf{r}_2)\mathbf{A}_{12}\cdot(\mathbf{p}_1 - \mathbf{p}_2)\Psi(\mathbf{r}_2)\Psi(\mathbf{r}_1), \tag{5.70}$$

$$H_{2b} = \tfrac{1}{2}\iint d\mathbf{r}_1 d\mathbf{r}_2 \Psi^+(\mathbf{r}_1)\Psi^+(\mathbf{r}_2) v(\mathbf{r}_1,\mathbf{r}_2)\Psi(\mathbf{r}_2)\Psi(\mathbf{r}_1), \tag{5.71}$$

$$\begin{aligned} H_3 = \tfrac{1}{6}\iiint & d\mathbf{r}_1 d\mathbf{r}_2 d\mathbf{r}_3 \Psi^+(\mathbf{r}_1)\Psi^+(\mathbf{r}_2)\Psi^+(\mathbf{r}_3) \\ & \times w(\mathbf{r}_1,\mathbf{r}_2,\mathbf{r}_3)\Psi(\mathbf{r}_3)\Psi(\mathbf{r}_2)\Psi(\mathbf{r}_1). \end{aligned} \tag{5.72}$$

The factors $\frac{1}{2}$ and $\frac{1}{6}$ result from the symmetrization of the interaction potentials v and w with respect to integration variables:

$$v(\mathbf{r}_1,\mathbf{r}_2) = \frac{e^2}{mc^2}|\mathbf{A}_{12}|^2, \tag{5.73}$$

$$w(\mathbf{r}_1,\mathbf{r}_2,\mathbf{r}_3) = \frac{e^2}{mc^2}(\mathbf{A}_{12}\cdot\mathbf{A}_{13} + \mathbf{A}_{21}\cdot\mathbf{A}_{23} + \mathbf{A}_{31}\cdot\mathbf{A}_{32}). \tag{5.74}$$

When only the two-particle interaction is considered, the motion equation for the Green functions has the form (Fetter 1992):

$$\left(i\hbar\frac{\partial}{\partial t_1} + \frac{\hbar\nabla_1^2}{2m}\right)G(1,1') + i\int d\mathbf{r}_2 v(\mathbf{r}_1 - \mathbf{r}_2)G_2(12,1'2^+)|_{t_2=t_1} = \hbar\delta(1-1'), \tag{5.75}$$

where the following notation has been introduced: $1 = \mathbf{r}_1 t_1$, and 2^+ means that the time t_2' is infinitesimally bigger than t_2 and $\mathbf{r}_2' = \mathbf{r}_2$. The appropriate Green functions are defined as follows

$$G(1,1') = -i\langle T[\Psi(1)\Psi^+(1')]\rangle, \tag{5.76}$$

$$G(12,1'2') = (-i)^2\langle T[\Psi(1)\Psi(2)\Psi^+(2')\Psi^+(1')]\rangle. \tag{5.77}$$

On the other hand, in the case of the Chern–Simons system, the three-particle interactions should be taken into account, and then the equation of motion assumes the form:

$$\begin{aligned}&\left(i\hbar\frac{\partial}{\partial t_1} + \frac{\hbar\nabla_1^2}{2m}\right)G(1,1') + i\frac{e}{mc}\int d\mathbf{r}_2[\mathbf{A}_{12}\cdot(\mathbf{p}_1 - \mathbf{p}_2)]G_2(12,1'2')|_{2'=2^+,t_2=t_1}\\ &+ i\int d\mathbf{r}_2 v(\mathbf{r}_1 - \mathbf{r}_2)G_2(12,1'2^+)|_{t_2=t_1}\\ &+ \tfrac{1}{2}\iint d\mathbf{r}_2 d\mathbf{r}_3 w(\mathbf{r}_1,\mathbf{r}_2,\mathbf{r}_3)G_3(123,1'2^+3^+)|_{t_3=t_2=t_3} = \hbar\delta(1-1'),\end{aligned} \tag{5.78}$$

where

$$G_3(123,1'2'3') = (-i)^3\langle T[\Psi(1)\Psi(2)\Psi(3)\Psi^+(3')\Psi^+(2')\Psi^+(1')]\rangle. \tag{5.79}$$

In the Hartree approximation, it is assumed that

$$G_2(12,1'2') \approx G(1,1')G(2,2') \tag{5.80}$$

and

$$G_3(123,1'2'3') \approx G(1,1')G(2,2')G(3,3'). \tag{5.81}$$

The irreducible self-energy Σ (mass operator) is defined by the following equation:

$$\left(i\hbar\frac{\partial}{\partial t_1} + \frac{\hbar\nabla_1^2}{2m}\right)G(1,1') - \int d2\hbar\Sigma(1,2)G(2,1') = \hbar\delta(1-1'). \tag{5.82}$$

In the Hartree expression for self-energy, a divergent term from the short-range interaction H_{2b} appears. This term is omitted in the present case because when the appropriate exchange (Fock) term is taken into account, this divergence disappears. Then,

$$\hbar\Sigma_{\mathrm{H}}(3,4) = \hbar\Sigma_{\mathrm{H}}(\mathbf{r}_3,\mathbf{r}_4)\delta(t_3 - t_4), \tag{5.83}$$

where

$$\begin{aligned}\hbar\Sigma_{\mathrm{H}}(\mathbf{r}_3,\mathbf{r}_4) = &-\frac{1}{2}\delta(\mathbf{r}_3 - \mathbf{r}_4)\iint d\mathbf{r}_7 d\mathbf{r}_8 w(\mathbf{r}_1,\mathbf{r}_2,\mathbf{r}_3)G(7,7^+)G(8,8^+)\\ &- i\frac{e}{mc}\iint d\mathbf{r}_7 d\mathbf{r}_8[\mathbf{A}_{78}\cdot(\mathbf{p}_7 - \mathbf{p}_8)G(8,8')\\ &\times\delta(\mathbf{r}_4 - \mathbf{r}_7)\delta(\mathbf{r}_3 - \mathbf{r}_7')]|_{\mathbf{r}_7'=\mathbf{r}_7,8'=8^+}.\end{aligned} \tag{5.84}$$

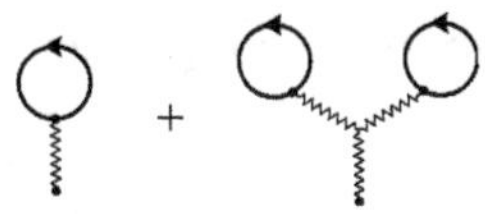

FIG. 5.2. Feynman diagrams of the self-energy of the system of anyons in the Hartree approximation.

Figure 5.2 illustrates, in a simplified form, contributions to the irreducible self-energy within the Hartree approximation. One can easily prove that the energy defined in this way, leads exactly to single-particle eigenstates as it also obtained in the mean field approximation (Fetter and Hanna 1992*a*) (Appendix B).

Baym and Kadanoff proved that the conservation rules (of a particle number, an energy, a momentum, an angular momentum) are satisfied if a functional derivative of the irreducible self-energy over the self-consistent Green function satisfies the following relation (Baym and Kadanoff 1961; Kadanoff and Baym 1962):

$$\frac{\delta\Sigma(1,1')}{\delta G(2',2)} = \frac{\delta\Sigma(2,2')}{\delta G(1',1)}. \tag{5.85}$$

This means that the self-energy can be obtained from the functional Ψ according to the following expression:

$$\hbar\Sigma(1,1') = \frac{\delta\Psi}{\delta G(1',1)}. \tag{5.86}$$

Baym and Kadanoff obtained the linear integral equation for the two-particle correlation function F:

$$F(11',2'2) = iG_2(12,1'^{+}2'^{+}) - iG(1,1'^{+})G(2,2'+). \tag{5.87}$$

Let F_0 be the two-particle unperturbed correlation function:

$$F_0(11',2'2) = -iG(1,2'^{+})G(2,1'^{+}). \tag{5.88}$$

The integral kernel (vertex function) of the equation for the correlation function F is determined by the variational derivative:

$$W(34,56) = i\frac{\delta\hbar\Sigma(3,4)}{\delta G(5,6)}\bigg|_{t_4=t_3^{+},t_6=t_5^{+}}. \tag{5.89}$$

Thus, the appropriate (Bethe–Salpeter) equation is (Fetter 1992)

$$F(11',2'2) = F_0(11',2'2) + \iiint d\mathbf{r}_3 d\mathbf{r}_4 dt_3 \iiint d\mathbf{r}_5 d\mathbf{r}_6 dt_5 F_0(11',34)W(34,56)F(56,2'2). \tag{5.90}$$

If Σ is determined by the functional Ψ, then the obtained approximated characteristics of transport phenomena will satisfy the conservation rules of a number of particles, a momentum, an energy, and an angular momentum (for the electromagnetic perturbation, the response will be gauge invariant). In the present case, W will assume the form

$$W_{\rm H}(34,56) = W_{\rm H}(\mathbf{r}_3\mathbf{r}_4,\mathbf{r}_5\mathbf{r}_6)\delta(t_3 - t_5), \tag{5.91}$$

where

$$\begin{aligned} W_{\rm H}(\mathbf{r}_3\mathbf{r}_4,\mathbf{r}_5\mathbf{r}_6) = &-i\int d\mathbf{r}_7 w(\mathbf{r}_3,\mathbf{r}_5,\mathbf{r}_7)G(7,7^+)\delta(\mathbf{r}_3-\mathbf{r}_4)\delta(\mathbf{r}_5-\mathbf{r}_6) \\ &+ \frac{e}{mc}\iint d\mathbf{r}_7 d\mathbf{r}_8 [\mathbf{A}_{78}\cdot(\mathbf{p}_7-\mathbf{p}_8) \\ &\times \delta(\mathbf{r}_5-\mathbf{r}_8)\delta(\mathbf{r}_6-\mathbf{r}'_8)\delta(\mathbf{r}_4-\mathbf{r}_7)\delta(\mathbf{r}_3-\mathbf{r}'_7)]|_{\mathbf{r}'_7=\mathbf{r}_7,\mathbf{r}'_8=\mathbf{r}_8}, \end{aligned} \tag{5.92}$$

which is shown, in the simplified form in Fig. 5.3. In the three-particle interaction term, one variable is integrated out which results in the appearance of the two-particle effective interaction (logarithmic). It can be proved that the self-consistent Hartree approximation and the RPA, discussed earlier (assuming gauge invariance), result in the same correlation functions (Fetter and Hanna 1992*a*) (Appendix B).

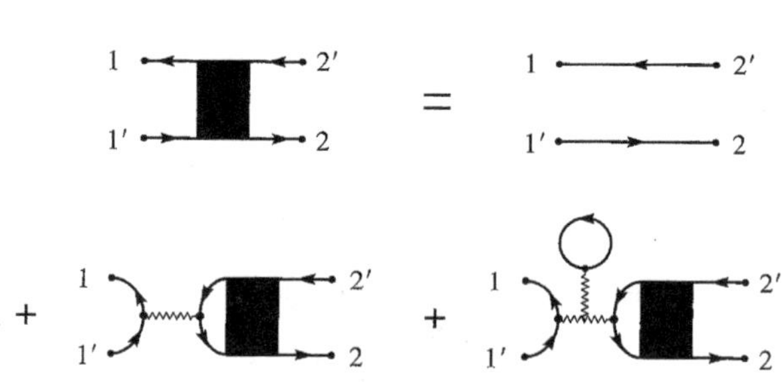

FIG. 5.3. Feynman diagrams defining the two-particle Green function for anyons in the self-consistent Hartree approximation.

5.4.2 *Self-consistent Hartree–Fock approximation for gas of anyons*

In the Hartree–Fock approximation, the eigenstates become anti-symmetrized. The two-particle Green function should be approximated in the following way:

$$G_2(12, 1'2') \approx G(1,1')G(2,2') - G(1,2')G(2,1'). \tag{5.93}$$

However, the three-particle Green function can be expressed as follows:

$$\begin{aligned} G_3(123, 1'2'3') \approx\ & G(1,1')G(2,2')G(3,3') - G(1,1')G(2,3')G(3,2') \\ & + G(1,2')G(2,3')G(3,1') - G(1,2')G(2,1')G(3,3') \\ & + G(1,3')G(2,1')G(3,2') - G(1,3')G(2,2')G(3,1'). \end{aligned} \tag{5.94}$$

Using the equation of motion (5.78) and the definition of the irreducible self-energy (5.82), the self-energy in the Hartree–Fock approximation is obtained, which is shown in a simplified form in Fig. 5.4. They correspond to the equations obtained in Section 5.3 (Dai *et al.* 1992). The appropriate Hartree–Fock mass

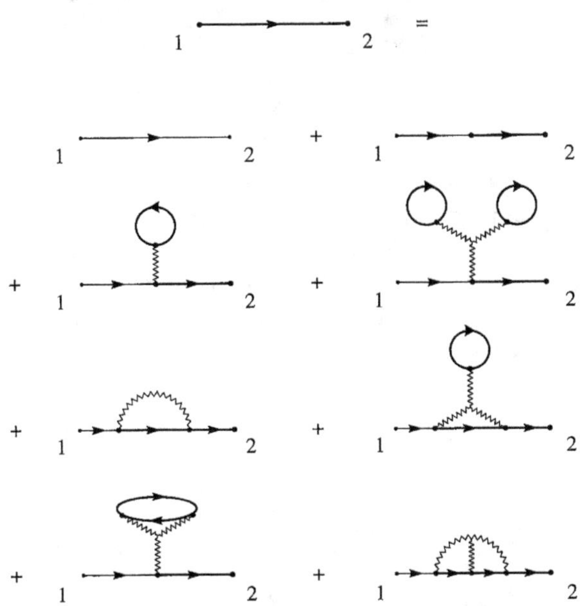

FIG. 5.4. Feynman diagrams defining the one-particle Green function for the system of anyons in the Hartree–Fock approximation.

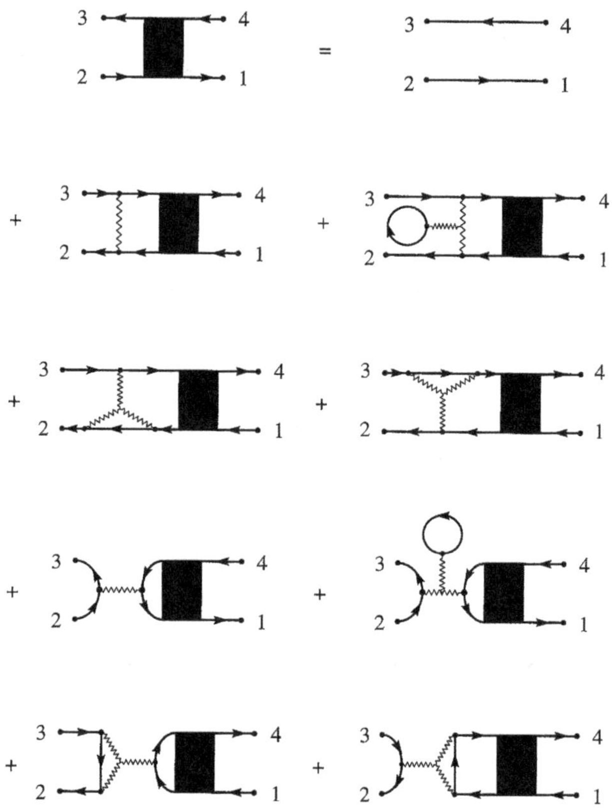

FIG. 5.5. Feynman diagrams defining the two-particle Green function in the self-consistent Hartree–Fock approximation.

operator is introduced in F_0 and the vertex function is determined, as in the case of the Hartree approximation. Then, the Bethe–Salpeter equation is illustrated in the simplified form in Fig. 5.5. Such approximation has been presented in the work (Dai *et al.* 1992), to which we will often refer to because in the case of anyons it leads to the most precise results.

6

ANYON SUPERCONDUCTIVITY

6.1 Meissner effect in an anyon gas at $T = 0$

Let us consider a gas of anyons in the absence of an external magnetic field when $\pi/(\pi - \theta) = f$ is an integer. In the fermion representation there are f first Landau levels (in the mean field approach) fully occupied[1] which results in the appearance of an energy gap in the spectrum of single-particle excitations, as in the case of superconducting systems. Anyon superconductivity was first suggested by Laughlin for *semions*, that is, anyons of the statistics parameter $\theta = \pi/2$ ($f = 2$) (Kalmeyer and Laughlin 1987; Laughlin 1988*a*; Wen and Zee 1990). At first, a model of paired semions was suggested as such a pair (at strong pairing limit) can be treated as a boson (in analogy to the fermion BCS superconductivity). As it will be shown in this chapter, the pairing of anyons is not necessary (although it can be considered—e.g. paired Hall states are interpreted in this way (Greiter *et al.* 1991; Moore and Read 1991)) for the appearance of superconductivity. Anyon superconductivity is rather similar to superconductivity of a system of charged bosons (Chen *et al.* 1989; Fetter *et al.* 1989).

For the sake of notation simplicity, in this section $\omega_c = eB^s/cm$ denotes the frequency unit and $a_0 = (\sqrt{eB^s/\hbar c})^{-1}$ denotes the length unit. At temperature $T = 0$, the correlation function $D_0^{\mu\nu}$ of the undisturbed system is given as (Jacak and Sitko 1993) (Appendix B):

$$D_R^0(\mathbf{q}, \omega) = \frac{f}{2\pi}\begin{pmatrix} q^2\Sigma_0 & -iq\Sigma_1 \\ iq\Sigma_1 & \Sigma_2 \end{pmatrix}, \tag{6.1}$$

where

$$\Sigma_j = \frac{e^{-x}}{f}\sum_{m=f}^{\infty}\sum_{l=0}^{f-1}\frac{m-l}{(\omega+i\eta)^2-(m-l)^2}\frac{l!}{m!}x^{m-l-1}[L_l^{m-l}(x)]^{2-j}$$
$$\times\left[(m-l-x)L_l^{m-l}(x) + 2x\frac{dL_l^{m-l}(x)}{dx}\right]^j, \tag{6.2}$$

[1] A flux attached to a particle is then equal to Φ_0/f, where $\Phi_0 = hc/e$ denotes the quantum of magnetic flux. The degeneration of Landau levels equals the number of magnetic flux quanta coming through the system. As f particles carry one flux quantum there are, therefore, f Landau levels fully occupied.

$x = \mathbf{q}^2/2$. The response matrix is as follows (in units of $(4\pi\rho e^2/mc^2)$)

$$K^{\text{RPA}} = \frac{1}{W}\begin{pmatrix} q^2\Sigma_0 & -iq(\Sigma_s - \Sigma_0) \\ iq(\Sigma_s - \Sigma_0) & 1 + \Sigma_2 \end{pmatrix}, \tag{6.3}$$

where $W = \det(I - D^0 V) = (1 + \Sigma_1)^2 - \Sigma_0(1 + \Sigma_2)$, $\Sigma_s = \Sigma_1 + \Sigma_1^2 - \Sigma_0\Sigma_2$. The response matrix K is gauge invariant, which can be tested by examining the response to a gauge field (Dai *et al.* 1992) (this expression is connected with an appropriate Ward identity (Enz 1992)):

$$K\begin{pmatrix} -\omega \\ q \end{pmatrix} = 0. \tag{6.4}$$

In the self-consistent Hartree and Hartree–Fock approximations, the gauge invariance is obtained automatically.

To describe the response in the limits $q \to 0$, $\omega \to 0$, we consider series expansions of Σ_j $(x = q^2/2)$:

$$\Sigma_0 \simeq -1 - \omega^2 + \tfrac{3}{4} f x, \tag{6.5}$$

$$\Sigma_1 \simeq -1 - \omega^2 + \tfrac{3}{2} f x, \tag{6.6}$$

$$\Sigma_2 \simeq -1 - \omega^2 + 2 f x. \tag{6.7}$$

Considering the gauge $\nabla \cdot \mathbf{A} = 0$ and $\mathbf{q} = q\hat{\mathbf{x}}$, the response to a magnetic field is given by the response matrix element K^{yy} (Fetter and Walecka 1971). The following equation is obtained (Fetter *et al.* 1989):

$$K^{yy}(\mathbf{q}, 0) \xrightarrow{q \to 0} 1, \tag{6.8}$$

which means the presence of the Meissner effect[2] in the system of anyons for the statistics parameter $\theta = \pi(1 - 1/f)$.

The response function poles, which define collective excitations, can be obtained via finding zeros of the determinant W. Further calculations lead to a sequence of excitations shifted by $\omega_c : \omega_m(q \to 0) = m\omega_c$, $m = 2, 3, \ldots$ (Hanna and Fetter 1993; Lopez and Fradkin 1993) (Fig. 6.1). The lowest excitation has a phonon dispersion relation $(q \to 0)$

$$\omega_1 \simeq \sqrt{f} q, \tag{6.9}$$

[2] As $A^x_{\mathbf{q}} = 0$, then $K^{yy} = \text{const}$ denotes that $J^y = \text{const} \cdot A^y$. Thus, a current and a vector potential appear to be proportional, like in the London equation, which results in the Meissner effect.

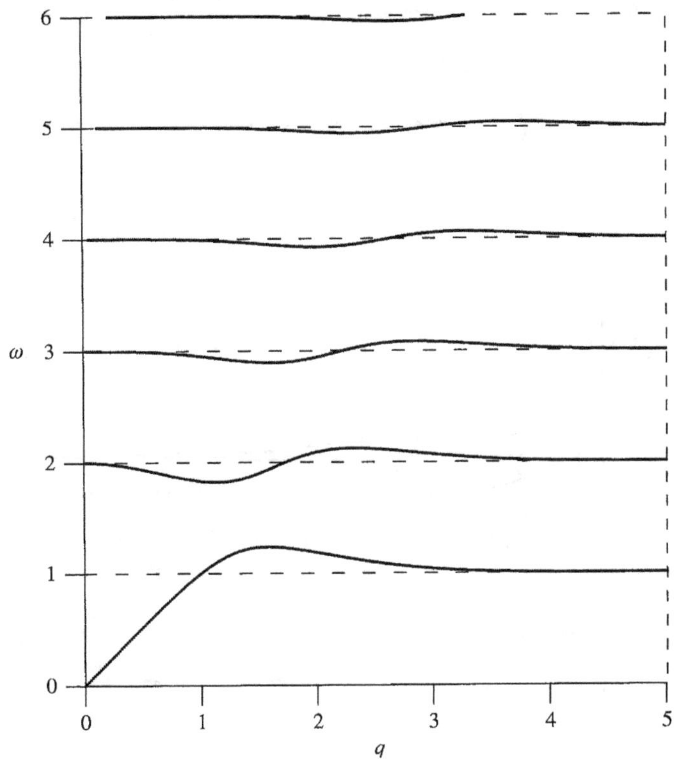

FIG. 6.1. Collective excitations of bosons in the fermion representation in the RPA.

which is also characteristic of superfluid systems.[3] For $f = 2$, a roton minimum can be observed (Hanna and Fetter 1993) (Fig. 6.2). According to Kohn's theorem, in a system in an external magnetic field, a sequence of excitations with gaps given by a multiplicity of cyclotron frequency should appear. The introduction of the 'statistical' mean field generates similar modes; however, the phonon mode, found in the random access approximation (RPA), shows that the interaction H_{int} crucially modifies the mean field picture (Lopez and Fradkin 1993). The mean field approach breaks the commutation rule $[p_x, p_y] \neq 0$, therefore, the gapless phonon mode can be treated as a Goldstone mode (Chen *et al.* 1989).

[3]If a body of a mass M (energy $E = Mv^2/2$, momentum $\mathbf{P} = M\mathbf{v}$) excites in a superfluid liquid, a quasiparticle $(\epsilon,\mathbf{q})$, then:

$$E = E' + \epsilon, \quad M\mathbf{v} = M\mathbf{v}' + \mathbf{q}.$$

If such $c > 0$ exists that for each $\mathbf{q}$, $\epsilon \geq cq$, then the above equations (momentum and energy conservation rules) can be solved provided that $v \geq c$. For $v < c$, a quasiparticle cannot be excited, therefore, the body does not change the energy and momentum.

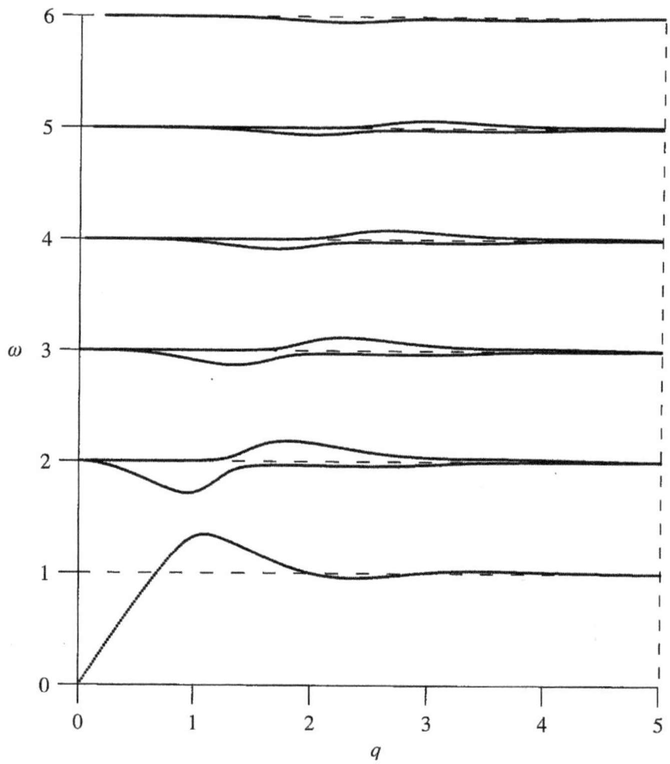

FIG. 6.2. Collective excitations of semions in the fermion representation in the RPA.

Let us consider now off-diagonal elements of the matrix $K^{\mu\nu}$ which constitute the Hall conductivity. We have

$$\sigma_{xy}(q,\omega) = \frac{1}{i\omega}K^{xy} = \frac{f}{2\pi W}(\Sigma_0 - \Sigma_s) \simeq |_{q\to 0,\, \omega\to 0}\, \frac{f}{8\pi}\,\frac{fq^2}{\omega^2 - fq^2}. \tag{6.10}$$

One can see that $\sigma_{xy}(q \to 0, 0)$ is non-zero, hence a Hall effect is found.[4] Since there is no external magnetic field, the Hall effect is an artefact of our approximation. Within the self-consistent Hartree–Fock approximation (Dai *et al.* 1992), the Hall effect disappears for bosons; however, it is still present for anyons of the statistics parameter $\theta = \pi(1 - 1/f)$ (Dai *et al.* 1992). So that, one can expect that within a similar model the fractional quantum Hall effect can be described—and in fact that is achieved within the composite fermion approach (see Chapter 7).

[4] When the electric field is only in y direction ($E_x = 0$), the Hall current is $J^x = \sigma_{xy}E_y$.

6.2 Gas of anyons at finite temperatures

The anyon gas at non-zero temperatures can be described using the Green–Matsubara functions. Let us first consider a typical sum over Matsubara frequencies present in the Matsubara correlation function (RPA bubble graph—see Fig. 5.1)

$$d_{ij}(u_k) = \frac{1}{\beta\hbar}\sum_{\omega_l}\frac{1}{[i\omega_l + iu_k - ((\epsilon_i-\mu)/\hbar)]}\frac{1}{[i\omega_l - ((\epsilon_j-\mu)/\hbar)]}, \tag{6.11}$$

where $\epsilon_j = (j+\frac{1}{2})\hbar\omega_c$, $\omega_l = ((2l+1)/\hbar\beta)\pi$, and $u_k = (2k/(\hbar\beta))\pi$ is the Matsubara energy transfer for fermions (l, k are integer numbers). Assuming that $u_k \neq 0$, one gets the known result (Fetter and Walecka 1971):

$$d_{ij}(u_k) = \frac{f_j - f_i}{iu_k - (i-j)\omega_c}, \tag{6.12}$$

where $f_j = (\exp(\epsilon_j - \mu)\beta + 1)^{-1}$.

However, due to the fact that within the mean field approach the Fourier transform does not lead to quantum numbers,[5] especially important becomes the term for $u_k = 0$ and $i = j$—then one gets a double pole in d_{jj} and the integration over residua gives the following expression[6] (Randjbar-Daemi *et al.* 1990; Hetrick *et al.* 1991; Fetter and Hanna 1992*a*)

$$d_{jj}(u_k = 0) = -\beta\hbar f_j(1-f_j). \tag{6.13}$$

As we shall see this result has the crucial effect on results at non-zero temperatures in the RPA. The correlation function of unperturbed system is given by the same matrix (6.1), however, ($iu_k \to \omega + i\eta$) (Jacak and Sitko 1993)

$$\Sigma_j(\omega\neq 0) = \frac{e^{-x}}{f}\sum_{m=1}^{\infty}\sum_{l=0}^{m-1}(f_l - f_m)\frac{m-l}{((\omega/\omega_c)+i\eta)^2-(m-l)^2}\frac{l!}{m!}x^{m-l-1}$$
$$\times [L_l^{m-l}(x)]^{2-j}\left[(m-l-x)L_l^{m-l}(x) + 2x\frac{dL_l^{m-l}(x)}{dx}\right]^j. \tag{6.14}$$

[5]In the zeroth order approximation, the Chern–Simons field is replaced by the mean field leading to breaking of the commutation rule $[p_x, p_y] \neq 0$. It means that the Fourier variable (momentum) is not a good quantum number.

[6]In the case of a double pole in $f(z)$, the integral is (Rudin 1974):

$$\mathrm{res}_{z_0} f = \lim_{z\to z_0}\frac{d}{dz}[(z-z_0)^2 f(z)].$$

In our case we deal with the derivative over the Fermi–Dirac distribution function, leading to the result (6.13).

For $\omega = 0$ one gets the additional term which equals (Fetter and Hanna 1992*a*):

$$\Sigma_j(\omega = 0) = \Sigma_j(\omega \to 0) - \frac{\beta\hbar\omega_c e^{-x}}{2fx} \times \sum_m [L_m(x)]^{2-j} \left[-xL_m(x) + 2x\frac{dL_m(x)}{dx}\right]^j f_m(1 - f_m). \quad (6.15)$$

Note that the temperature-dependent correlation function (5.21) is not a continuous function at $\omega = 0$.

In order to find collective excitations, let us find the series expansions of Σ_j in q and ω:

$$\Sigma_0 \simeq -1 - \omega^2 + \tfrac{3}{4}\alpha x, \quad (6.16)$$

$$\Sigma_1 \simeq -1 - \omega^2 + \tfrac{3}{2}\alpha x, \quad (6.17)$$

$$\Sigma_2 \simeq -1 - \omega^2 + 2\alpha x, \quad (6.18)$$

where

$$\alpha = 2\frac{1}{f}\sum_{m=0}^{\infty} f_m \left(m + \frac{1}{2}\right). \quad (6.19)$$

The chemical potential is given by the condition $\sum f_m = f$, hence at low temperatures ($\beta\hbar\omega_c \gg 1$) one gets (Randjbar-Daemi *et al.* 1990; Fetter and Hanna 1992*a*):

$$\mu \simeq f - \frac{1}{2\beta\hbar\omega_c}\exp(-f\beta\hbar\omega_c). \quad (6.20)$$

Then, at $T \to 0$, the following approximated expression can be found:

$$\alpha \simeq f + \frac{2}{f}\exp\left(-\frac{\beta\omega_c}{2}\right). \quad (6.21)$$

We can see that the collective excitations spectrum is similar to the $T = 0$ case—the lowest mode is the gapless mode

$$\omega_1 \simeq \sqrt{\alpha}q. \quad (6.22)$$

However, for $\omega = 0$, the series expansions of Σ_j are as follows:

$$\Sigma_0(q,0) \simeq -\frac{\tau_0}{2x} - 1 + \frac{1}{2}(\tau_0 + 2\tau_1) + \frac{1}{4}x[3\alpha - (\tau_0 + 3\tau_1 + 3\tau_2)], \quad (6.23)$$

$$\Sigma_1(q,0) \simeq -1 + \tfrac{1}{2}(\tau_0 + 2\tau_1) + \tfrac{1}{2}x[3\alpha - (\tau_0 + 3\tau_1 + 3\tau_2)], \quad (6.24)$$

$$\Sigma_2(q,0) \simeq -1 + 2x\left[\alpha - \tfrac{1}{4}(\tau_0 + 4\tau_1 + 4\tau_2)\right], \quad (6.25)$$

where $\tau_j = \beta\hbar\omega_c/f \sum_{m=0}^{\infty} m^j f_m(1-f_m)$. Using these formulas one can find the response to a static magnetic field, that is, the term K^{yy} of the response matrix—related to the Meissner effect (Jacak and Sitko 1993) ($q \to 0$):

$$K^{yy}(q,0) \simeq \frac{q^2}{\tau_0 + q^2}, \tag{6.26}$$

and at low temperatures (Fetter and Hanna 1992*a*)

$$\tau_0 \simeq \frac{2\beta\hbar\omega_c}{f} \exp\left(-\frac{\beta\hbar\omega_c}{2}\right). \tag{6.27}$$

Hence, within the RPA the Meissner effect disappears at finite temperatures ($T \neq 0$), $K^{yy}(q \to 0, 0) = 0$. However, one can speculate that this is an artefact of our approximation. Within the self-consistent Hartree–Fock approximation, it is shown that the response function does not depend on temperature and the Meissner effect is present at all temperatures[7] (Dai *et al.* 1992).

6.3 Higgs mechanism in an anyon superconductor

Anyon superconductivity, as discussed in the previous paragraphs, at $T = 0$ differs crucially from BCS superconductivity. Note that the Meissner effect (6.8) was found in the system with no Coulomb interaction. The question arises whether an inclusion of a Coulomb interaction leads to a Higgs mechanism and is connected with a Meissner effect.[8] As we will see the result depends on electrodynamics being considered. In real systems one should rather consider three-dimensional electrodynamics and electrostatic potential $v_{3D} = 2\pi e^2/\epsilon q$ (two-dimensional Fourier transform of the potential $\sim 1/r$). Purely theoretically, it is also interesting to consider the electrostatic potential of two-dimensional electrodynamics, $v_{2D} = 2\pi e^2/\epsilon q^2$, obtained from the Poisson equation in two-dimensional.[9]

[7] If a superconducting state can be treated as a Kosterlitz–Thouless phase (Dai *et al.* 1992), then the transition temperature—at which the correlated vortex–antivortex pairs are destroyed—is related to a vortex–antivortex interaction (logarithmic). In our case a vortex–antivortex pair is a particle-hole pair (above Fermi sea—filled Landau levels)—a gap in the single-particle excitation spectrum is related to a particle–hole interaction.

Within the simple RPA, single-particle gaps do not depend on the vortex–antivortex interaction. It is not supprising then that a transition from a superconducting to a normal state takes place at any non-zero temperatures. In the case of the self-consistent Hartree–Fock approximation, as considered in Dai *et al.* (1992), the discussed interaction is so strong (a single-particle excitation gap is divergent with a system size) that it leads to an infinite transition temperature to a normal state (Kosterlitz–Thouless transition). Then, both approximations give unphysical results. In Kitazawa and Murayama (1990), the phenomenological method of finding a finite transition temperature (and of inclusion of interaction) is given.

[8] If superconductivity is interpreted in terms of a spontaneous symmetry breaking (gauge symmetry), then a gap in the collective excitations spectrum is connected with the Higgs mechanism in which a gap corresponds to an excitation mass (Weinberg 1986).

[9] Here, we deal with major mathematical difficulties when solving the Poisson equation in two-dimension. The main solution is a logarithmic function. In the Fourier picture one gets $\sim 1/q^2$, but only in the sense of the principal value of the integral (Vladimirov 1981).

As it is known, the Higgs mechanism in condensed matter physics is related to the presence of a gap in a spectrum of collective excitations of a charged system (i.e. in the presence of a long-ranged gauge field and a spontaneous symmetry breaking). In the BCS theory of superconductivity, this effect is related to the Meissner effect (Weinberg 1986) and is often called the dressing of a photon (with a mass).[10]

Considering the electrostatic potential v_{3D} one gets the following dispersion of the lowest collective excitation of interacting anyons (Bujkiewicz *et al.* 1994) (Fig. 6.3):

$$\omega_1^2 \simeq f r_s q + \alpha q^2, \tag{6.28}$$

where we have introduced the dimensionless parameter $r_s = a_0(e^2 m/\epsilon\hbar^2) = a_0/a_B$, a_B is the Bohr radius. However, the Meissner kernel at temperature $T = 0$ behaves as for the non-interacting system, $K_{3D}^{yy}(q \to 0, 0) \to 1$, at non-zero

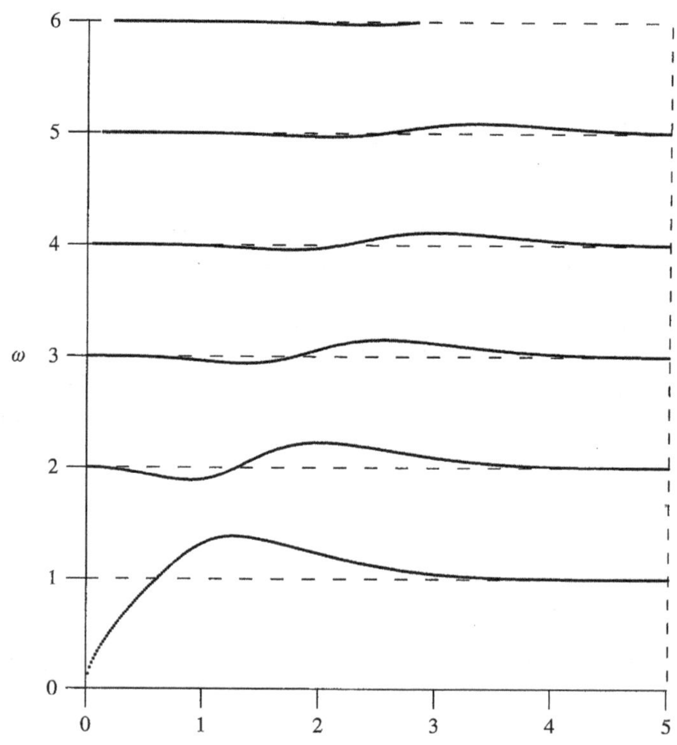

FIG. 6.3. Collective excitations of Coulomb interacting bosons for v_{3D} in the fermion representation in the RPA.

[10]Putting the London equation $\mathbf{J} = \text{const} \cdot \mathbf{A}$ into the Maxwell equation, one gets massive electromagnetic waves (Weinberg 1986).

temperatures

$$K^{yy}_{3\mathrm{D}} \simeq \frac{fr_{\mathrm{s}}\gamma q^2}{\tau_0(\gamma q + fr_{\mathrm{s}})} \xrightarrow{q\to 0} 0, \tag{6.29}$$

where $\gamma = [-1 + \frac{1}{2}(\tau_0 + 2\tau_1)]^2 + \tau_0[\alpha - \frac{1}{4}(\tau_0 + 4\tau_1 + 4\tau_2)]$. Hence, despite the introduction of the Coulomb interaction there is no gap in the collective excitation spectrum—the dispersion relation is as for a two-dimensional plasmon ($\omega \sim \sqrt{q}$). If strictly two-dimensional electrodynamics is considered, the electrostatic potential is $v_{2\mathrm{D}}$ and (Bujkiewicz *et al.* 1994)

$$\omega_1^2 = \frac{fr_{\mathrm{s}}}{1 - fr_{\mathrm{s}}} + \alpha\frac{1 - (3/8fr_{\mathrm{s}})}{1 - fr_{\mathrm{s}}}q^2, \tag{6.30}$$

(see Figs 6.4 and 6.5) Eqn (6.30) is valid only for small gaps ($\omega \ll \omega_{\mathrm{c}}$). When $\omega \sim \omega_{\mathrm{c}}$, the gap of the excitation ω_1 is given by $\omega_1(q \to 0) = \sqrt{fr_{\mathrm{s}}}$. The presence of the Coulomb interaction results in a gap in the collective

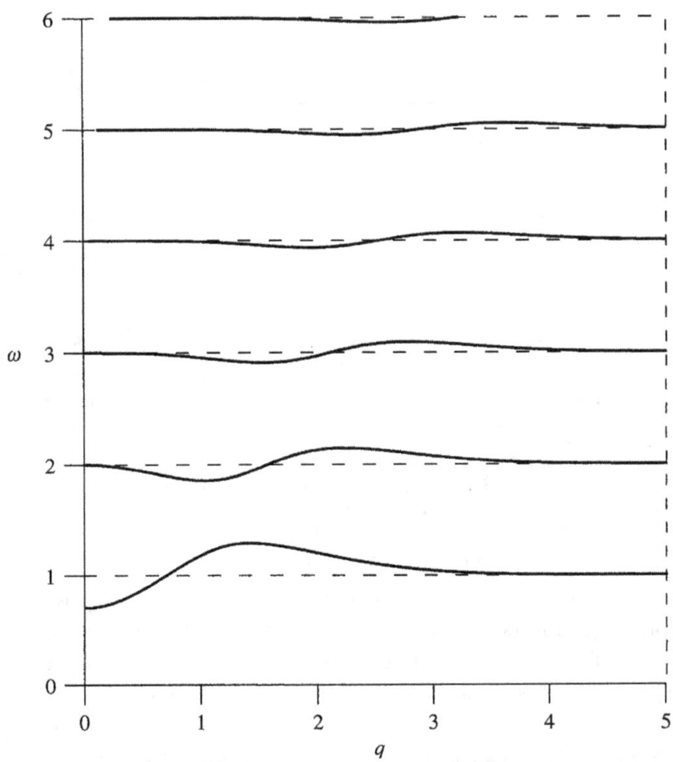

FIG. 6.4. Collective excitations of Coulomb interacting bosons for $v_{2\mathrm{D}}$ ($r_{\mathrm{s}} = 0.5$) in the fermion representation in the RPA.

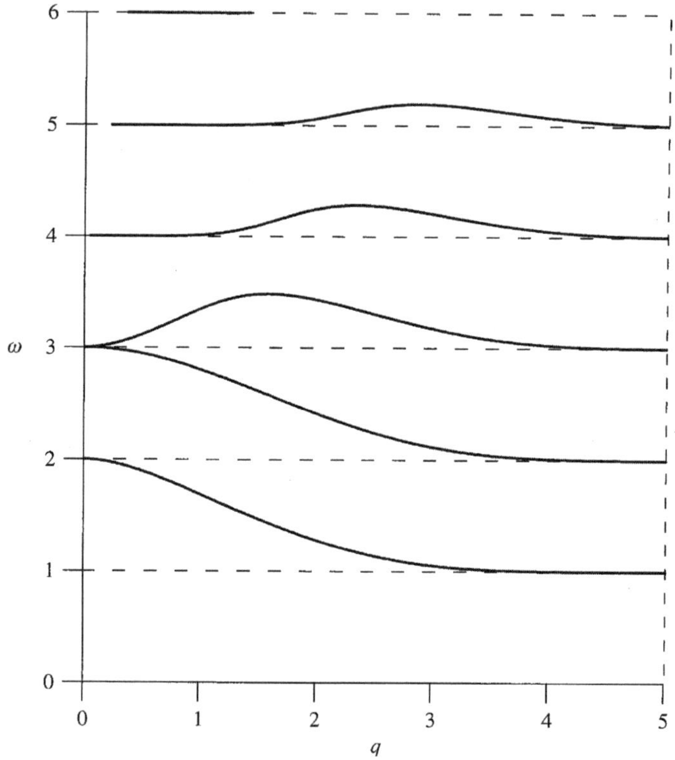

FIG. 6.5. Collective excitations of Coulomb interacting bosons for v_{2D} $(r_s = 9.0)$ in the fermion representation in the RPA.

excitations spectrum—the Higgs mechanism[11]—which appears to be independent of the dimension of the system, keeping in mind that proper electrodynamics is considered. The Meissner effect for v_{2D} disappears at $T \neq 0$ (RPA)

$$K^{yy}_{2D} \simeq \frac{f r_s \gamma q^2}{\tau_0(\gamma q^2 + f r_s)} \xrightarrow{q \to 0} 0, \tag{6.31}$$

but is present at $T = 0$: $K^{yy}_{2D}(q \to 0, 0) \to 1$.

As it is seen, the Higgs mechanism in anyon systems is present only when two-dimensional Coulomb interaction is considered. There is no correspondence to the Meissner effect, however, and this is one of the differences between the BCS superconductivity and the anyon superconductivity.

[11]A spontaneous symmetry breaking results in the appearance of a Goldstone boson. In the system of non-interacting anyons, this is the phonon mode $\omega \simeq \sqrt{f}q$. In the Coulomb interacting system (long-ranged gauge field), the boson is 'dressed' with a mass what is called a Higgs mechanism.

Considering collective modes in the system of interacting anyons at finite temperatures (6.28, 6.30), especially the Higgs mechanism for v_{2D}, one reaches the conclusion that the temperature-dependent response function is right for $\omega \neq 0$, $q \to 0$. For non-interacting systems and for the interactions v_{2D} and v_{3D} $(T \neq 0)$:

$$K^{yy}(q \to 0, \omega \to 0) = 1, \tag{6.32}$$

it seems to be right to assume that the Meissner effect is present at non-zero temperatures. As it was mentioned, that assumption was confirmed in the self-consistent Hartree–Fock approximation (Dai *et al.* 1992).

6.4 Ground state of an anyon superconductor in the Hartree–Fock approximation

The ground state of an anyon superconductor within the Hartree–Fock approximation was first found for bosons ($\theta = 0$) and semions ($\theta = \pi/2$) (Laughlin 1988*b*; Hanna *et al.* 1989), the extension of those results for anyons of statistics parameters $\theta = \pi(1 - 1/f)$ was given in (Sitko 1992) and (Fetter and Hanna 1992*b*). Here we present only an analysis of the final results. The Hamiltonian of a single particle in the Hartree–Fock approximation for $\theta = \pi(1 - 1/f)$, given in terms of projection operators (on the kth Landau level) Π_k,[12] is given by (in units of $\hbar\omega_c$):

$$\begin{aligned} H_{\mathrm{HF}} = & \sum_{k=0}^{f-1} \left(k + \frac{1}{2} - \frac{1}{2f} - \frac{1}{2f} E_R + \frac{1}{4f} S_{f-1} - \frac{1}{2f} S_{f-1-k} \right) \Pi_k \\ & + \sum_{k=f}^{\infty} \left[k + \frac{1}{2} - \frac{1}{4f^2} + \frac{1}{2f} E_R + \frac{1}{4} \left(\frac{1}{k} + \frac{1}{k+1} \right) \right. \\ & \left. - \frac{1}{4f} S_{f-1} + \frac{1}{2f} S_{k-f} \right] \Pi_k, \end{aligned} \tag{6.33}$$

where $E_R = \ln Ra_0^{-1} + \frac{1}{2}(\gamma - \ln 2)$ ($\gamma \simeq 0.577\ldots$ is the Euler constant), R is the sample radius, $S_m = \sum_{k=1}^{m} \frac{1}{k}$, and $S_0 = 0$.

The important result is that in the Hartree–Fock approximation, the single-particle Hamiltonian can be written in terms of Landau level projection operators. It means that the starting (mean field) wave function is a good choice. The first sum in the above expression is over filled Landau levels, the second sum is over empty ones. One can see that the divergent terms (E_R) appear with opposite signs. It means that the energy gap between the filled Landau levels and the next ones is of the order of E_R/f—which is infinite in the thermodynamic limit ($R \to \infty$). The gap is generated by the exchange term of the logarithmic interaction H_{HF}^{12}, however, if this interaction is screened (e.g. takes a form $1/(q^2 + a)$

[12] $\Pi_k = \sum_l |\phi_{kl}\rangle\langle|\phi_{kl}|$, where ϕ_{kl} is the wave function of a particle on the kth Landau level, l numbers orbital state within the Landau level (degenerated).

in the Fourier image), the divergent gap disappears. Then it is important to find out whether the interaction can be screened.

Let us consider the effective interaction within the RPA:

$$V^{\mathrm{RPA}} = V[I - D^0 V]^{-1} = \frac{2\pi}{fq^2 W} \begin{pmatrix} 1+\Sigma_2 & iq(1+\Sigma_1) \\ -iq(1+\Sigma_1) & q^2\Sigma_0 \end{pmatrix}. \tag{6.34}$$

Since $1 + \Sigma_2(q \to 0, 0) = O(q^2)$ and $W(q \to 0, 0) = O(q^2)$, the logarithmic interaction remains unscreened (of the order of $1/q^2$) (Dai *et al.* 1992). Then, the divergent gap in the single-particle spectrum is not modified within the RPA. In an analogy to the Kosterlitz–Thouless problem, one can treat a single-particle excitation as a vortex and a hole as an antivortex. This analogy is supported by more precise calculations of the two-particle Green function within the Bethe–Salpeter equation (Dai *et al.* 1992), where the divergence in Hartree–Fock energies cancels out the divergence in the vertex function (Hanna *et al.* 1991). Because of the logarithmic attraction between a hole and a particle, a binding energy is infinite, and the correlation function of an anyon superconductor does not depend on temperature, the Meissner effect is present at all temperatures. We can add that Kitazawa and Murayama (1990) found a Kosterlitz–Thouless transition temperature within a phenomenological model, however, their results are not confirmed within the self-consistent Hartree–Fock approximation (Dai *et al.* 1992).

The ground-state energy of the system within the Hartree–Fock approximation is finite:

$$\langle H \rangle = \frac{1}{2} N E_{\mathrm{F}} \left(1 - \frac{1}{4f^2} - \frac{1}{4f^3}\right), \tag{6.35}$$

which also confirms the choice of the unperturbed ground state as the ground state of the mean field Hamiltonian ($E_{\mathrm{F}} = 2\pi\hbar^2\rho/m$—the Fermi energy of the two-dimensional electron gas). It is interesting to consider the Hartree–Fock result for the known case—bosons. The free bosons system should condense in the state of the lowest energy, that is, in the zero momentum $\mathbf{p} = 0$ and zero energy state. The energy of bosons ($f = 1$) found from Eqn (6.35) is non-zero; however, it is half of the mean field result. The lower value is found considering the correlation energy within the RPA (Hanna and Fetter 1993). This example shows that a proper picture of bosons within the fermion representation is practically impossible, by considering higher classes of diagrams one gets closer to good results, though.

7

THE FRACTIONAL QUANTUM HALL EFFECT IN COMPOSITE FERMION SYSTEMS

7.1 Hall conductivity in a system of composite fermions

The basic idea of composite fermions was conceived by Jain and closely resembled Laughlin's idea of the trial wave function. The Laughlin wave function can be split up in the following way:

$$\Psi_{1/2p+1} = \prod_{i<j}(z_i - z_j)^{2p}\chi_1, \tag{7.1}$$

where χ_1 denotes a wave function of a system with the lowest Landau level filled up, the factor denotes the Jastrov factor. One should notice that the magnetic field, present in χ_1 refers to a real magnetic field. When particles are transposed, the Jastrov factor introduces an additional phase which can be interpreted as an effect of attaching of fluxes to electrons (i.e. even multiplicity of magnetic flux quantum). Then, while calculating the effective magnetic flux per particle, the actual filling of the lowest Landau level is determined, that is, $1/(2p+1)$ (this can also be interpreted as a result of a strong repulsion present in the Jastrov factor). Jain suggested an analogous trial function for other fillings (Jain 1989)

$$\Psi_{n/2pn\pm1} = \prod_{i<j}(z_i - z_j)^{2p}\chi_n, \tag{7.2}$$

where the filling fraction can be easily determined because χ_n denotes the wave function of the system of n Landau levels filled up completely. Many numerical studies on finite systems of few particles (Dev and Jain 1992; Jain and Kawamura 1995) proved that the trial function (7.2) constitutes (with accuracy of 99 per cent) the ground state of a system of few electrons, interacting through Coulomb forces, in very strong magnetic fields. To obtain the above-mentioned comparison, the wave function (7.2) was projected on the lowest Landau level as the initial system exists on the lowest Landau level (Dev and Jain 1992).

The equivalent definition of composite fermions was proposed by Greiter and Wilczek (however they used another name—*superfermions*) (Greiter and Wilczek 1990). In that approach, ordinary fermions carrying point magnetic fluxes ($2p$ quanta of magnetic flux are attached to a single fermion) are called composite fermions. The considerations on anyons (Chapters 5 and 6) suggest that such a system can be described within the Chern–Simons theory (Chen *et al.* 1989). The Chern–Simons composite fermion approach has been developed in the works of Lopez and Fradkin (1991, 1992, 1993). Within the mean field approximation, the Chern–Simons model is equivalent to the Laughlin–Jain approach. However, determining Laughlin or Jain functions on the basis of the Chern–Simons theory is not so obvious (for the Laughlin wave function it is achieved within the lowest Landau-level theory—see Shankar and Murthy (1997; Murthy and Shankar 1998*b*)). In the Chern–Simons theory, the wave function of a system of composite fermions is

$$\Psi = \prod_{i<j} \frac{(z_i - z_j)^{2p}}{|z_i - z_j|^{2p}} \chi, \tag{7.3}$$

where the magnetic field which appears in the function Ψ is identical to the one in the function χ, and the particle density does not change because $|\Psi|^2 = |\chi|^2$. The phase factor in the above formula can be introduced into the Hamiltonian (as a result of a gauge transformation), which leads to the Chern–Simons field (similarly as in the case of anyons). Lopez and Fradkin pointed out that the Laughlin (Jain) wave function refers rather to a cut-off Chern–Simons interaction, as it takes place in the random phase approximation (RPA) (Lopez and Fradkin 1992). However, how the two approaches are related to each other has not been fully explained yet. On the other hand, it is the Chern–Simons theory which has been successfully used by Halperin, Lee, and Read (HLR) to define characteristics of a metallic state of composite fermions in good agreement with the experimental data (Halperin *et al.* 1993; Willett *et al.* 1993; Du *et al.* 1994; Manoharan *et al.* 1994).

In the following discussion, the system of composite fermions will be analysed within the Chern–Simons field theory. Composite fermions are defined by means of the statistics parameters $\theta = (1+2p)\pi$, for p being an integer. They differ from ordinary fermions by an additional phase $p \cdot 2\pi$ when the particles are transposed. The case when there are n completely filled Landau levels in the effective field B^* (the statistical field is opposite to the external field $B^{\mathrm{s}} = -2p(hc/e)\bar{\rho}$) is the subject of the present analysis:

$$B^* = B^{\mathrm{s}} + B^{\mathrm{ex}} = \frac{1}{n}\frac{hc}{e}\bar{\rho}. \tag{7.4}$$

In reference to the real external magnetic field, it means that the lowest Landau level is filled in the part $\nu = n/(2pn+1)$.[1] Within the RPA, the response matrix is given by $(\hbar, c = 1,\ \omega_c^{\mathrm{eff}} = eB^*/m)$ (Sitko 1994*a*):

$$K^{\mathrm{RPA}} = \frac{e^2 n}{2\pi W} \begin{pmatrix} \frac{q^2}{\omega_c^{\mathrm{eff}}}\Sigma_0 & -iq(\Sigma_s + 2pn\Sigma_0) \\ iq(\Sigma_s + 2pn\Sigma_0) & (1+\Sigma_2)\omega_c^{\mathrm{eff}} \end{pmatrix}, \tag{7.5}$$

where $W = \det(I - D^0 V) = (1 - 2pn\Sigma_1)^2 - (2pn)^2\Sigma_0(1+\Sigma_2)$, $\Sigma_s = \Sigma_1 - 2pn\Sigma_1^2 + 2pn\Sigma_0\Sigma_2$.

[1] A case when the effective field is opposite to the external field can also be considered; then $\nu = n/(2pn - 1)$.

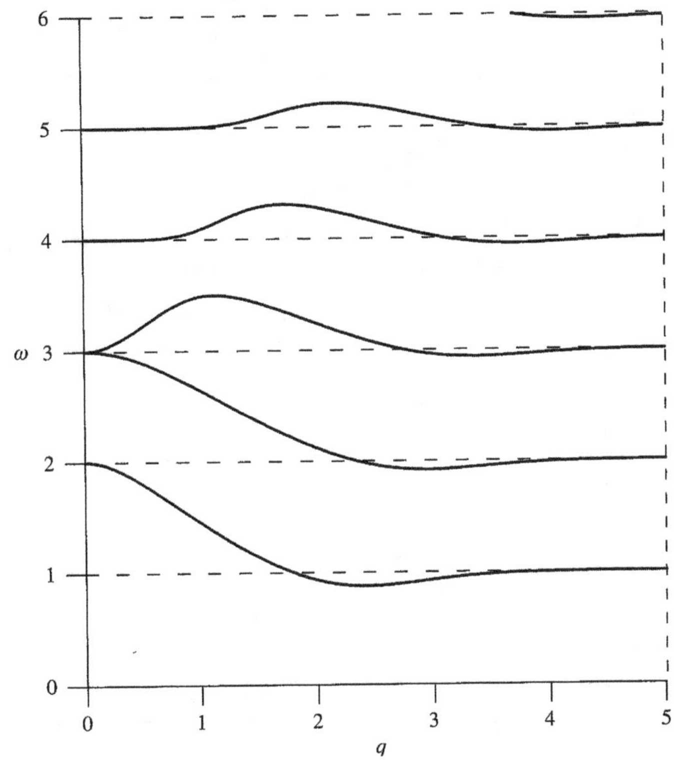

FIG. 7.1. Collective excitations in a system of composite fermions at the filling 1/3.

While studying collective excitations (i.e. $W = 0$), it is found that there are a number of excitations with energy gaps $\omega_m(q \to 0) = m\omega_c^{\rm eff}$, $m = 2, 3, \ldots$ (Lopez and Fradkin 1993; Sitko and Jacak 1994). In particular, numerical characteristics of excitations for $\nu = 1/3, 1/5, 2/5$ (Sitko and Jacak 1994) are shown in Figs 7.1–7.3. They agree with Lopez and Fradkin's (1993) suggestion that there is a split-up excitation for $\omega = \omega_c^{\rm ex}$. Consequently, in a system of composite fermions, beside excitations related to the effective field (which seem to result from the approximation inadequacies), there also exist excitations with gaps corresponding to the external magnetic field (Lopez and Fradkin 1993). When $n = 1$, double poles in the expression $\Sigma_1^2 - \Sigma_0\Sigma_2$ cancel out. However, when $n = 2$, double poles are present and cause (unlike Lopez and Fradkin's (1993) result, but in accord with anyon analysis (Hanna and Fetter 1993)) an additional splitting of each branch of excitation: $\omega_m^-(q \to \infty) = m\omega_c^{\rm eff} = \omega_m^+(q \to \infty)$, $m > 1$ (look at Fig. 7.3). The lowest excitations in Figs 7.1 and 7.3 were confirmed in the recent resonant inelastic light scattering experiments of Kang *et al.* (2000, 2001).

At temperature $T = 0$, the system does not exhibit a Meissner effect (i.e $K^{yy}(\omega = 0, q \to 0) = 0$) (Sitko 1994*a*), and the Hall (transverse) conductivity is

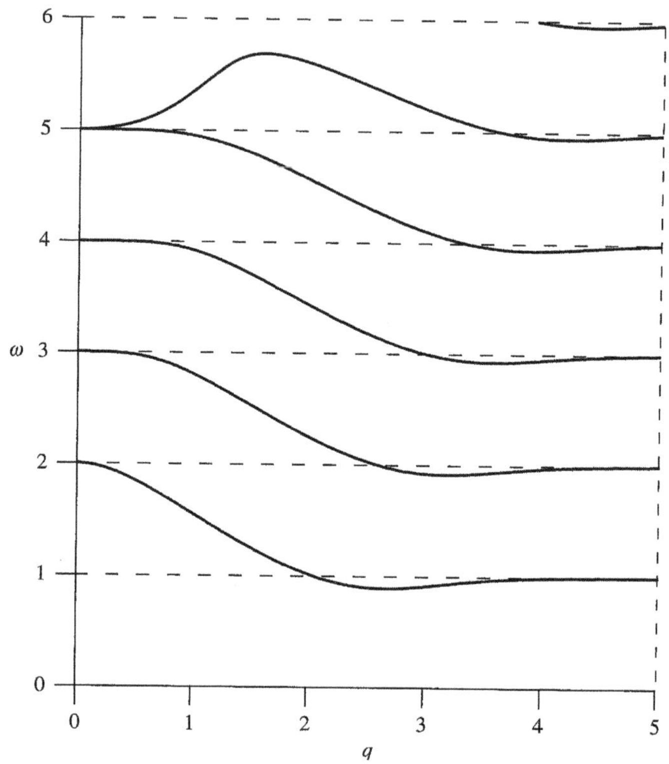

FIG. 7.2. Collective excitations in a system of composite fermions at the filling 1/5.

equal to (Fradkin 1991; Lopez and Fradkin 1991; Sitko 1994*a*)

$$\sigma_{xy}(q,0) = \frac{e^2 n}{2\pi W}(-2pn\Sigma_0 - \Sigma_s) \simeq |_{q\to 0}\frac{e^2}{2\pi}\frac{n}{1+2pn} = \frac{e^2}{2\pi}\nu, \tag{7.6}$$

which constitutes a fraction of the value e^2/h. Thus, the value of the Hall conductivity, at the filling $\nu = n/(2pn+1)$, is exactly as it is observed in the fractional quantum Hall effect (FQHE) (Tsui *et al.* 1982).

The temperature dependence of the longitudinal resistance ρ_{xx} constitutes one of the arguments for the correctness of composite fermion theory (Du *et al.* 1993, 1994; Manoharan *et al.* 1994). Let us note that in the integer quantum Hall effect (IQHE) $\rho_{xx} \sim \exp(-\Delta/2kT)$, where the experimentally determined energy gap Δ value is very close to the cyclotron energy $\hbar\omega_c^{\rm ex}$ (Prange and Girvin 1987). The Hall resistance $\rho_{xy}(T)$ also depends weakly on temperature, which is defined by the following expression: $\rho_{xy}(T) - \rho_{xy}(0) = \Delta\rho_{xy}(T) = -s\rho_{xx}(T)$, where $s = 0.01 \div 0.5$ (Prange and Girvin 1987). A similar dependence of ρ_{xx} is found in the region of the FQHE, and moreover the energy gap corresponds to the mean field theory of composite fermions (Du *et al.* 1993, 1994; Manoharan

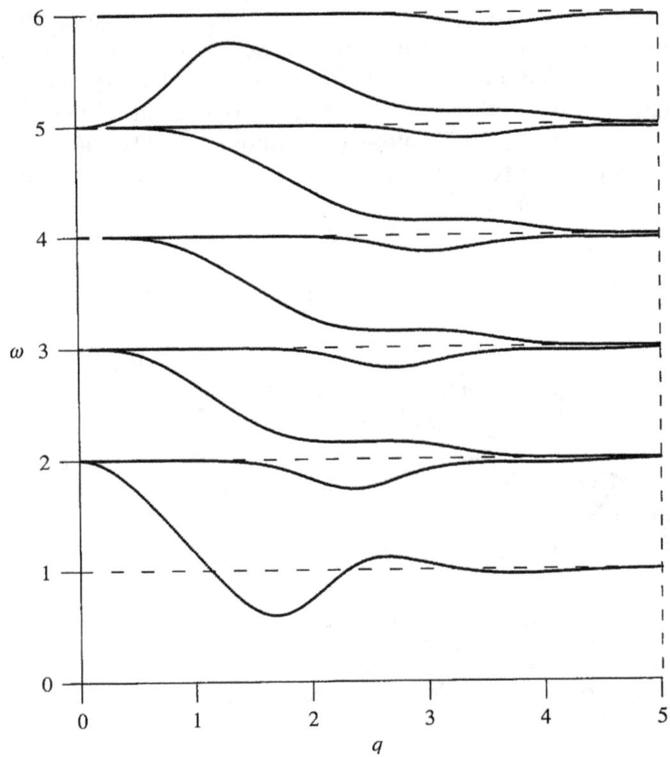

FIG. 7.3. Collective excitations in a system of composite fermions at the filling 2/5.

et al. 1994). We now present the analysis of temperature dependence of the linear response of composite fermion systems in the RPA. Introducing the Coulomb interaction v_{3D} (a more realistic one—see section 6.3), the following expression is obtained:

$$\sigma_{xy}(q,0) = \frac{e^2 n}{2\pi W}(-2pn\Sigma_0 - \Sigma_s) \overset{q\to 0}{\simeq} \frac{e^2 n}{2\pi} \times \frac{2pn\Omega_1^2 - \Omega_1 + 2pn\Omega_2}{(1-2pn\Omega_1)^2 + (2pn)^2\Omega_2 + nm/(2\pi)\tau_0 v_{3D}(q)}, \tag{7.7}$$

where $\Omega_1 = \frac{1}{2}(\tau_0 + 2\tau_1) - 1$, $\Omega_2 = \tau_0[\alpha - \frac{1}{4}(\tau_0 + 4\tau_1 + 4\tau_2)]$. From analogous analysis for anyons at low temperatures, the following approximated expressions can be used:

$$\tau_1 \simeq \frac{\beta\omega_c}{n}(2n-1)\exp\left(-\frac{\beta\omega_c}{2}\right), \tag{7.8}$$

$$\tau_2 \simeq \frac{\beta\omega_c}{n}[n^2 + (n-1)^2]\exp\left(-\frac{\beta\omega_c}{2}\right). \tag{7.9}$$

Eqn (7.7) shows that in a system with Coulomb interaction, the Hall effect disappears at non-zero temperatures. Let us note, however, that the numerical analysis (see Chapters 8 and 9) of systems of few particles indicates that a two-dimensional system of Coulomb interacting electrons can be treated as a gas of weakly interacting composite fermions (Dev and Jain 1992; Jain and Kawamura 1995). That is why a system with no Coulomb interaction is considered below. Then, at low temperature limit:

$$\sigma_{xy}(T) - \sigma_{xy}(0) \simeq -\frac{e^2}{2\pi}\frac{n}{(1+2pn)}2\beta\omega_{\rm c}\exp\left(-\frac{\beta\hbar\omega_{\rm c}}{2}\right). \tag{7.10}$$

Assuming that $\Delta\sigma_{xy}(T) \sim -\rho_{xx}(T)$, the above result, (7.10), agrees qualitatively with the experimental data (Du *et al.* 1993, 1994; Manoharan *et al.* 1994) if the electron mass m is substituted with the effective mass m^*.

It is also interesting to consider analogous results for the IQHE, that is, the system with no Chern–Simons field ($p = 0$). For a Coulomb interacting system the, Hall effect disappears at non-zero temperatures ($\sigma_{xy}(q \to 0, 0) = 0$). In a system with no Coulomb interaction, the following is found:

$$\sigma_{xy}(q \to 0, 0) = \frac{ne^2}{2\pi}(1 - \tau_0 - 2\tau_1), \tag{7.11}$$

which can be approximated at low temperatures as:

$$\Delta\sigma_{xy}(T) \simeq -\frac{n}{2\pi}4\beta\hbar\omega_{\rm c}^{\rm ex}\exp\left(-\frac{\beta\hbar\omega_{\rm c}^{\rm ex}}{2}\right). \tag{7.12}$$

This result roughly agrees with experiments (Prange and Girvin 1987), although it is not clear why in the case of the IQHE the correct result has been obtained with no Coulomb interaction included. It is likely that taking into account the effects of both impurities and disorder, the relation between the theory and the experiment will be found.

7.2 Ground-state energy of composite fermion systems

The Hartree–Fock approximation for a system of composite fermions in the FQHE leads to the following expression for the effective single-particle Hamiltonian $H_{\rm HF}$ (Sitko 1994*a*) (in units of $\hbar\omega_{\rm c}^{\rm eff}$):

$$\begin{aligned} H_{\rm HF} = {} & \sum_{k=0}^{n-1}\left(k+\frac{1}{2}+2p+2p^2n-2p^2nE_{\rm R}+p^2nS_{n-1}-2p^2nS_{n-1-k}\right)\Pi_k \\ & + \sum_{k=n}^{\infty}(k+\frac{1}{2}-p^2+2p^2nE_{\rm R}+p^2n^2\left(\frac{1}{k}+\frac{1}{k+1}\right)-p^2nS_{n-1} \\ & + 2p^2nS_{k-n})\Pi_k. \end{aligned} \tag{7.13}$$

Thus, an infinite (in thermodynamic limit, compare with Eqn (6.4)) energy gap has again been obtained. The gap appears between the last filled Landau level and the next empty one.

As in the case of anyons, one should consider the effective interaction in the RPA:

$$\begin{aligned} V^{\mathrm{RPA}} &= V[I - D^0 V]^{-1} \\ &= \frac{4p\pi}{q^2 W} \begin{pmatrix} 2pn\omega_c^{\mathrm{eff}}(1+\Sigma_2) & -iq(1-2pn\Sigma_1) \\ iq(1-2pn\Sigma_1) & 2pn\frac{q^2}{\omega_c^{\mathrm{eff}}}\Sigma_0 \end{pmatrix}. \end{aligned} \tag{7.14}$$

However, in the present case $W(q \to 0, 0) = (1+2pn)^2 + O(q^2)$, and the logarithmic interaction becomes screened off, which constitutes a basic difference between the system of composite fermions under consideration now, in the case of the FQHE, and an anyon superconductor. That is why an infinite energy gap in the ground state of the system of composite fermions in the Hartree–Fock approximation seems to artificially result from the approximation. The screening leads to a gap of a finite value (its value has only been estimated (Khveshchenko 1994)), which is in accord with experimental results (Kang *et al.* 1993; Willett *et al.* 1993; Goldman *et al.* 1994). Within the lowest Landau-level theory (Read 1994; Shankar and Murthy 1997; Halperin and Stern 1998; Murthy and Shankar 1998*a*; Pasquier and Haldane 1998) (in which flux and charge are separated in so-called dipolar composite fermions) the results for the energy gap are close to the numerical results (Murthy and Shankar 1999).

On the other hand, the average energy of the ground state equals (Sitko 1994*a*):

$$\langle H \rangle = \frac{1}{2} N E_{\mathrm{F}} \left(1 + 3p^2 + \frac{2p}{n} - \frac{p^2}{n}\right). \tag{7.15}$$

If the energy (7.15) is compared with the initial value of the energy of electrons in the external magnetic field $B^{\mathrm{ex}} = ((2pn+1)/n)(hc/e)\rho$ (the Coulomb interaction has been neglected), $\frac{1}{2}NE_{\mathrm{F}}(2p + (1/n))$, then the energy needed to transmute the system of electrons into the system of composite fermions is equal to:

$$\Delta E = \frac{1}{2} N E_{\mathrm{F}} \left[2p^2 + (p-1)^2 \left(1 - \frac{1}{n}\right)\right]. \tag{7.16}$$

The energy (7.16) differs from zero. It seems natural to assume that different realizations of fermion statistics should be equally allowed, and that is why the result (7.16) appears to be so unexpected. To verify result (7.16), the correlation energy of a system of composite fermions in the RPA was studied (Sitko and Jacak 1994). Let us remind that the correlation energy is defined as the correction of the ground-state energy with respect to the result obtained in the Hartree–Fock approximation:

$$E_{\mathrm{c}} = E - E_{\mathrm{HF}}. \tag{7.17}$$

On the basis of the earlier deduction (see Chapter 5), the correlation energy in the RPA, in the case of composite fermions, is expressed by the following equation (an analogous expression is found in the theory of a Fermi liquid (Fetter and Walecka 1971)):

$$E_{\mathrm{c}}^{\mathrm{RPA}} = -\frac{1}{2}\hbar\frac{N}{\rho}\int\frac{d\mathbf{q}}{(2\pi)^2}\int_0^\infty\frac{d\omega}{\pi}\int_0^1\frac{d\lambda}{\lambda} \times \mathrm{Im}\,\mathrm{Tr}\left\{\lambda V(q)[D_\lambda^{\mathrm{RPA}}(\mathbf{q},\omega) - D_0(\mathbf{q},\omega)]\right\}, \tag{7.18}$$

where D_λ^{RPA} denotes a time-ordered function (unlike, considered earlier, the retarded function) in the RPA with coupling constant λ:

$$D_\lambda^{\mathrm{RPA}}(\mathbf{q},\omega) = [I - \lambda D_0(\mathbf{q},\omega)V(\mathbf{q})]^{-1}D_0(\mathbf{q},\omega). \tag{7.19}$$

The results found in the work of Sitko and Jacak (1994) are presented now. Knowing the characteristics of collective excitations for $n = 1$, one obtains in $\hbar\omega_{\mathrm{c}}^{\mathrm{eff}}$ units (Hanna and Fetter 1993; Sitko and Jacak 1994):

$$\frac{E_{\mathrm{c}}^{\mathrm{RPA}}}{N} = \frac{1}{2}\sum_{m=1}^{\infty}\left(\int\Delta\omega_m(x)dx - (2p)^2\frac{1}{2m}\right), \tag{7.20}$$

where $\Delta\omega_m(x) = \omega_m(x) - m$. But for $n = 2$:

$$\frac{E_{\mathrm{c}}^{\mathrm{RPA}}}{N} = \frac{1}{4}\left(\int\Delta\omega_1(x)dx - 8p^2\right) + \frac{1}{4}\sum_{m=2}^{\infty}\left[\int\left(\Delta\omega_m^-(x) + \Delta\omega_m^+(x)\right)dx - (4p)^2\frac{1}{2m}\right] + p^2. \tag{7.21}$$

The results for numerical calculations for $\nu = 1/3, 1/5, 2/5$ are shown in Table 7.1. For $\nu = 1/3, 2/5$ (for $p = 1$), the transmutation energy is close to zero, but for $\nu = 1/5$ ($p = 2$) it is close to $(-E_{\mathrm{F}})$. It seems that the value for $\nu = 1/5$ is the result of a domination of the logarithmic interaction in the interaction Hamiltonian for $p > 1$, and after full consideration of three-particle

Table 7.1 *The correlation energy in the RPA at $\nu = 1/3,\ 1/5,\ 2/5$. In the value $\Delta E/N$, both the Hartree–Fock result and the RPA correlation energy are included.*

	1/3	1/5	2/5
$E_{\mathrm{c}}^{\mathrm{RPA}}/N$ in $\hbar\omega_{\mathrm{c}}^{\mathrm{eff}}$ units	−1.032	−5.274	−2.246
$\Delta E/N$ in E_{F} units	−0.032	−1.274	−0.123

interactions, the zero transmutation energy should be found. The results for Coulomb interacting systems can be found in Sitko (1999).

7.3 Metal of composite fermions

So far, we have considered only systems of composite fermions corresponding to FQHE at $\nu = n/(2pn+1)$. Both Laughlin's and Jain's ideas also referred to those fillings for which FQHE is observed. In 1993, HLR indicated that in areas where the quantum Hall effect had not appeared, composite fermions could still be realized. From the experimental point of view, the half-filling of the lowest Landau level, $\nu = 1/2$, appears to be extremely interesting. For the first generation of composite fermions ($p = 1$), the effective field at $\nu = 1/2$ equals zero, and a Fermi sea is a natural candidate for a ground state. The existence of such an unusual metallic state has been experimentally proved (Du *et al.* 1993; Kang *et al.* 1993; Willett *et al.* 1993) (Fig. 7.4). However, the characteristics of this new Fermi liquid are still not fully known. In RPA, HLR have shown that the effective mass becomes infinite at the Fermi momentum (Du *et al.* 1994; Manoharan *et al.* 1994). This puts in question the Fermi liquid nature of the ground

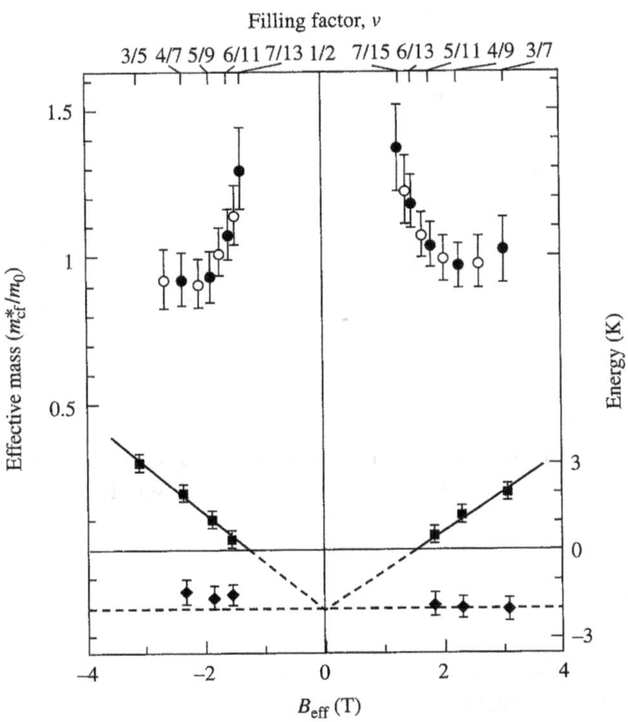

Fig. 7.4. The divergence of the effective mass of composite fermions (upper part, energy gaps are presented in the lower part) from Du *et al.* (1994).

state. Such divergence can be interpreted as a breaking of the quasiparticle model of the Fermi liquid (Simon and Halperin 1993, 1994; Nayak and Wilczek 1994). Varying the interaction between fermions $v(\mathbf{q}) = V_0/q^{2-\eta}$ $(1 \leq \eta \leq 2)$, the fluctuations of the gauge field are modified and the effective mass is changed. The fluctuations of the gauge field are found to be more singular for short-range interactions $(\eta \sim 2)$. Longer range interactions $(\eta \sim 1)$ prevent density from fluctuating. In the RPA, the self-energy $\Sigma(\mathbf{k}, \omega)$ behaves in the following way (Halperin *et al.* 1993):

$$\mathrm{Re}\,\Sigma \sim \mathrm{Im}\,\Sigma \sim \omega^{2/(1+\eta)} \quad \text{for } 1 < \eta \leq 2 \tag{7.22}$$

and

$$\mathrm{Re}\,\Sigma \sim \omega \ln \omega, \quad \mathrm{Im}\,\Sigma \sim \omega \qquad \text{for } \eta = 1. \tag{7.23}$$

Therefore, the Landau condition[2] is not fulfilled for quasiparticles for $1 < \eta \leq 2$. For $\eta = 1$ (Coulomb interaction) the system behaves as a marginal Fermi liquid (Kim *et al.* 1994). In both cases, the effective mass is divergent at the Fermi momentum:

$$\frac{m^*}{m} \sim |\xi_k|^{-(\eta-1)/(\eta+1)} \text{ dla } \quad 1 < \eta \leq 2, \tag{7.24}$$

$$\frac{m^*}{m} \sim |\ln \xi_k| \text{ dla } \quad \eta = 1, \tag{7.25}$$

where $\xi_k = (k^2/2m) - \mu$. One can note that the singularity in the self-energy can artificially result from an adopted gauge because the one-particle Green function is not gauge invariant. If, however, density–density and current–current correlation functions, which are gauge invariant, are considered, then divergences in the self-energy are cancelled out by analogous divergences in the vertex function (by virue of the Ward identities) (Altshuler *et al.* 1994; Kim *et al.* 1994). Consequently, the density–density correlation function behaves similarly as in the case of normal Fermi liquid.

Here, we present the results of the Hartree–Fock approximation for the metallic state of composite fermions. When the lowest Landau level is filled at $\nu = 1/(2p)$, the ground-state energy in the Hartree–Fock approximation is finite and equals (Sitko 1994*b*):

$$\langle H \rangle = N \frac{E_{\mathrm{F}}}{2} (1 + 3p^2), \tag{7.26}$$

which supports the choice of the single-particle ground state in the form of a plane wave. The result (7.26) agrees with the previous result (7.15) at the limit $n \to \infty$.

[2] If $|\mathrm{Im}\,\Sigma| \ll |\mathrm{Re}\,\Sigma|$, then a single-particle excitation can be treated as a stable quasiparticle, as it happens for a normal Fermi liquid.

The energy spectrum for a metal of composite fermions for $\nu = 1/(2p)$ is described by the following expressions (Sitko and Jacak 1995):

$$\begin{aligned}\xi_{\mathrm{HF}}(|\mathbf{k}| < k_{\mathrm{F}}) =& \frac{\mathbf{k}^2}{2m} + 3p^2 E_{\mathrm{F}} + p^2 E_{\mathrm{F}} \lim_{N\to\infty} \ln N \\ & - \frac{2p^2 E_{\mathrm{F}}}{\pi \rho_k} \sum_{\mathbf{q}\neq\mathbf{k}} \frac{1}{|\mathbf{k}-\mathbf{q}|^2},\end{aligned} \tag{7.27}$$

$$\begin{aligned}\xi_{\mathrm{HF}}(|\mathbf{k}| > k_{\mathrm{F}}) =& \frac{\mathbf{k}^2}{2m} + 2p^2 E_{\mathrm{F}} \left(\ln \frac{|\mathbf{k}|}{k_{\mathrm{F}}} + \frac{1}{2} + \frac{k_{\mathrm{F}}^2}{\mathbf{k}^2} \right) + p^2 E_{\mathrm{F}} \lim_{N\to\infty} \ln N \\ & - \frac{2p^2 E_{\mathrm{F}}}{\pi \rho_k} \sum_{\mathbf{q}\neq\mathbf{k}} \frac{1}{|\mathbf{k}-\mathbf{q}|^2},\end{aligned} \tag{7.28}$$

where $\rho_k = S/(2\pi\hbar)^2$, $\mathbf{k}$ is the particle momentum ($|\mathbf{k}| = k$). Let us consider the single-particle excitations spectrum for values of $\mathbf{k}$ close to the Fermi momentum; if $(k_{\mathrm{F}} - k') > 0$ and $(k'' - k_{\mathrm{F}}) > 0$ do not tend to zero, then the main contribution to the energy difference $\xi(k'') - \xi(k')$ is divergent:

$$\xi(k'') - \xi(k') > 2p^2 E_{\mathrm{F}} \lim_{\delta\to 0} \left(\ln \frac{k_F + k'}{\delta} + \ln \frac{k_F - k'}{\delta} \right). \tag{7.29}$$

In this way an infinite single-particle energy gap is obtained at the Fermi momentum. HLR have shown (taking the exchange term of the logarithmic interaction) that the Fermi group velocity equals zero, that is, the energy spectrum becomes flat at the Fermi momentum. Consequently, an infinite gap in the first order of perturbation theory leads to a flat spectrum in the RPA (which seems a little too strong a change as for perturbation theory calculations).

In a metal of composite fermions in the RPA, the Fermi group velocity disappears which means that the effective mass becomes infinite (Halperin *et al.* 1993). From Eqns (7.27) and (7.28), one can note that the only contributions to the Fermi group velocity in the Hartree–Fock approximation come from the term $\mathbf{k}^2/2m$ and the exchange term of the logarithmic interaction. As both those expressions have been included in the RPA (Halperin *et al.* 1993), the Hartree–Fock approximation does not contribute any corrections to the Fermi group velocity obtained in the RPA.

7.4 BCS (Bardeen–Cooper–Schrieffer) paired Hall state

The composite Fermi liquid raises a possibility of the BCS state of composite fermions. The BCS pairing of composite fermions was first proposed by Moore and Read (1991), and Greiter, Wen, Wilczek (GWW) (Greiter *et al.* 1991, 1992). Many theoretical and experimental studies confirmed that the 5/2 Hall state (Willett *et al.* 1987) (here we include spins, and the lowest Landau level is filled by electrons of opposite spins, the next level is half-filled) is the paired Hall state (Morf 1998; Bonesteel 1999; Pan *et al.* 1999*b*; Sitko 2001).

GWW noted that the H_1 (see Chapter 5) term of the Chern–Simons interaction (in the Fourier image, $B^* = 0$, the statistics parameter $\epsilon = (\theta - \pi)/\pi$):

$$V_{\mathbf{k},\mathbf{k}'} = 2\pi i \frac{\epsilon}{m} \frac{\mathbf{k} \times \mathbf{k}'}{|\mathbf{k} - \mathbf{k}'|^2} \tag{7.30}$$

can be attractive (Greiter *et al.* 1991). Let us write the BCS gap equation for the interaction (7.30):

$$\Delta_{\mathbf{k}} = -\frac{1}{2} \sum_{\mathbf{k}'} \frac{\Delta_{\mathbf{k}'}}{E_{\mathbf{k}'}} V_{\mathbf{k},\mathbf{k}'} \tag{7.31}$$

where $E_{\mathbf{k}} = \sqrt{|\Delta_{\mathbf{k}}|^2 + \xi_{\mathbf{k}}^2}$ ($\xi_{\mathbf{k}} = k^2/2m - k_{\mathrm{F}}^2/2m$). If the sum in (7.31) is replaced by an integral, then

$$\Delta_{\mathbf{k}} = -\frac{1}{2} \frac{1}{(2\pi)^2} \int d^2\mathbf{k}' \frac{\Delta_{\mathbf{k}'}}{E_{k'}} V_{\mathbf{k},\mathbf{k}'}. \tag{7.32}$$

For p-wave pairing, one has $\Delta_{\mathbf{k}} = |\Delta_{\mathbf{k}}| \exp(i\theta)$ and

$$|\Delta_{\mathbf{k}}| = \frac{1}{2} \frac{\epsilon}{m} \frac{1}{(2\pi)} \int d^2\mathbf{k}' \frac{|\Delta_{\mathbf{k}'}| \sin\theta}{E_{k'}} \frac{kk' \sin\theta}{k^2 + (k')^2 - 2kk' \cos\theta}. \tag{7.33}$$

Putting $\lambda = \frac{1}{2}((k/k') + (k'/k))$, and $\int_0^{2\pi} d\theta (\sin\theta)^2/(\lambda - \cos\theta) = 2\pi(\sqrt{\lambda^2 - 1} - \lambda)$, one gets ($x = k/k_{\mathrm{F}}$, $\Delta_x = |\Delta_{\mathbf{k}}|/E_{\mathrm{F}}$) (Greiter *et al.* 1992):

$$\Delta_x = \frac{\epsilon}{2} \left(\frac{1}{x} \int_0^x \frac{(x')^2 \Delta_{x'}}{\sqrt{\Delta_{x'}^2 + ((x')^2 - 1)^2}} dx' + x \int_x^\infty \frac{\Delta_{x'}}{\sqrt{\Delta_{x'}^2 + ((x')^2 - 1)^2}} dx' \right). \tag{7.34}$$

This equation can easily be solved for $\epsilon \ll 1$ (Greiter *et al.* 1992), the results for $\Delta_1 (0 < \epsilon < 1)$ are presented in Fig. 7.5 (Sitko 2001).

However, there is still the logarithmic interaction (two-body part of H_2) (Chen *et al.* 1989; Dai *et al.* 1992) of the form

$$V_{\mathbf{k},\mathbf{k}'} = 2\pi \frac{\epsilon^2}{2m} \frac{k_F^2}{|\mathbf{k} - \mathbf{k}'|^2}, \tag{7.35}$$

which is strongly pair breaking (singularities are found when solving the gap equation (7.31)).

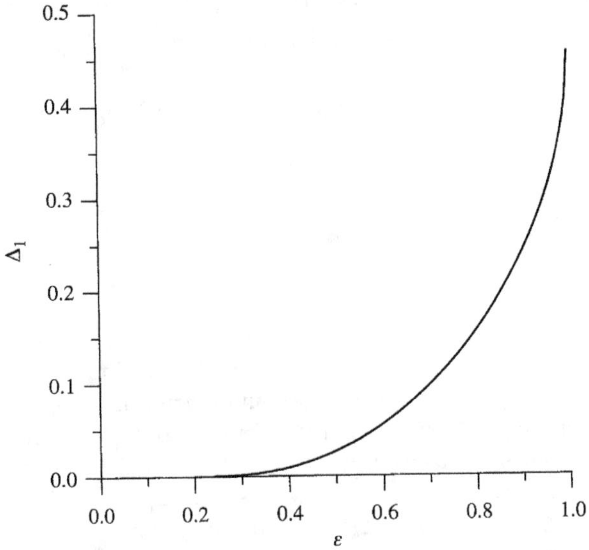

FIG. 7.5. The gap parameter $\Delta_1(\epsilon)$ for GWW interaction, $\epsilon = 1$ corresponds to bosons.

In order to avoid this difficulty, the screened RPA interaction, $(V^{\mathrm{sc}})^{-1} = V^{-1} - D^0$, has to be considered. We have

$$D^0(\mathbf{q}, \omega) = \begin{pmatrix} D_{00} & 0 \\ 0 & D_{yy} \end{pmatrix} \tag{7.36}$$

and for small q: $D^0_{00}(q,0) \simeq -m/2\pi$, $D^0_{yy}(q,0) \simeq -(\rho/m) + (q^2/12\pi m)$ $(q < 2k_{\mathrm{F}})$ (Halperin *et al.* 1993).

Dominating terms are

$$V_{\mathbf{k},\mathbf{k}'} = \frac{1}{q^2 m^2 \chi(q)} \left(i(\mathbf{k} \times \mathbf{k}') \frac{1}{\epsilon} + \frac{(\mathbf{k} \times \mathbf{k}')^2}{|\mathbf{k} - \mathbf{k}'|^2} \right), \tag{7.37}$$

where

$$q^2 \chi(q) = \frac{q^2 v(q)}{(2\pi\epsilon)^2} + \frac{(1 + (6/\epsilon^2))}{12\pi m} q^2. \tag{7.38}$$

We consider the case of $v(q)q^2 = \alpha q^{2-x} k_{\mathrm{F}}^x$ (ultra-long-range interactions $v(q) = \alpha k_{\mathrm{F}}^x / q^x$, $1 < x \leq 2$ (Halperin *et al.* 1993)) and for small q ($q^2\chi(q) \sim \frac{\alpha}{(2\pi\epsilon)^2} q^{2-x} k_{\mathrm{F}}^x$):

$$V_{\mathbf{k},\mathbf{k}'} = \frac{(2\pi\epsilon)^2}{\alpha} \frac{1}{m^2 k_{\mathrm{F}}^x} \left(\frac{i}{\epsilon} \frac{(\mathbf{k} \times \mathbf{k}')}{|\mathbf{k} - \mathbf{k}'|^{2-x}} + \frac{(\mathbf{k} \times \mathbf{k}')^2}{|\mathbf{k} - \mathbf{k}'|^{4-x}} \right). \tag{7.39}$$

Performing the integral over θ in the BCS gap Eqn (7.31) (taking $|\mathbf{k}| = |\mathbf{k}'| = k_{\mathrm{F}}$—Fermi-surface average) (Bonesteel 1999):

$$\lambda = -\frac{m}{2\pi}\int_0^{2\pi} V_{\mathbf{k},\mathbf{k}'} \exp{(i\theta)} d\theta, \quad |\mathbf{k}| = |\mathbf{k}'| = k_{\mathrm{F}} \tag{7.40}$$

one gets (Sitko 2001)

$$\lambda = 2^x \frac{2\pi}{\alpha m}\sqrt{\pi}\frac{\Gamma((x-1)/2)}{\Gamma((x+2)/2)}(2\epsilon - \epsilon^2), \tag{7.41}$$

where Γ is the Euler's gamma function. The constant λ is positive (attractive interaction) for $0 < \epsilon < 2$ and zero for $\epsilon = 0$ (fermions), $\epsilon = 2$ (composite fermions). Bonesteel (1999) showed that singularity in $V_{\mathbf{k},\mathbf{k}'}$ is strongly pair-breaking for short-range interactions (and for Coulomb interaction).

The analysis presented above supports the idea of the BCS paired Hall state (it is also supported by numerical studies—see Chapter 9). For higher Landau levels, that is, for the fillings $9/2, 11/2, \ldots$, the situation changes and the so-called stripe phase is observed (Pan *et al.* 1999*a*; Rezayi and Haldane 2000).

It should also be noted that quasiparticles (quasiholes) in the paired Hall Pfaffian (Pfaffian is the form of the BCS wave function in real space) state obey non-Abelian braid statistics (Moore and Read 1991; Nayak and Wilczek 1996; Read and Rezayi 1999) (see Section 3.3).

8

QUANTUM HALL SYSTEMS ON A SPHERE

8.1 Spherical system

A spherical geometry in the context of fractional quantum Hall effect (FQHE) was first introduced by Haldane (1983) in order to construct the hierarchy of FQHE states. In spite of the fact that the Haldane hierarchy picture is not commonly used today (we will comment on this further), the spherical geometry appeared to be a very convenient geometry for numerical studies of FQHE systems of a few electrons in a magnetic field. Forming a uniform magnetic field B piercing the sphere is the construction of a magnetic monopole. The magnetic flux is quantized (Dirac quantization) (Dirac 1931):

$$\Phi = 4\pi R^2 B = 2S\frac{hc}{e}, \tag{8.1}$$

where $2S$ is an integer, R is the the radius of the sphere. The single particle Hamiltonian can be written in the following way (Haldane 1983; Fano *et al.* 1986):

$$H_0 = \frac{|\mathbf{\Lambda}|^2}{2mR^2} = \frac{|\mathbf{\Lambda}|^2 eB}{2m\hbar Sc}, \tag{8.2}$$

where

$$\mathbf{\Lambda} = \mathbf{R} \times \left(-i\hbar\mathbf{\nabla} + \frac{e}{c}\mathbf{A}(\mathbf{R})\right) \tag{8.3}$$

and the vector potential of the monopole can be chosen as (in spherical coordinates):

$$\mathbf{A} = -\frac{\hbar Sc}{eR} ctg\theta \hat{\mathbf{n}}_\phi. \tag{8.4}$$

One can show (Fano *et al.* 1986) that $|\mathbf{\Lambda}|^2 = |\mathbf{L}|^2 - \hbar^2 S^2$, where $\mathbf{L} = \mathbf{\Lambda} + \hbar S\hat{\mathbf{\Omega}}$ is the angular momentum operator ($\hat{\mathbf{\Omega}} = \mathbf{R}/R, [L_\alpha, L_\beta] = i\hbar\epsilon_{\alpha\beta\gamma}L_\gamma$) and eigenvalues of $|\mathbf{L}|^2$ are $l(l+1)$. Hence, the eigenenergies of the Hamiltonian (8.2) are:

$$\xi_l = \frac{\hbar\omega_c}{2S}[l(l+1) - S^2], \tag{8.5}$$

$\omega_c = eB/mc$, $l = S, S+1, \ldots$. The eigenfunctions of an electron in a field of a monopole are given by so-called monopole harmonics (Wu and Yang 1976, 1977). The energy levels are Landau levels, in the present case they are angular momentum shells, the lowest shell is for $l = S$ ($\epsilon_S = \hbar\omega_c/2$).

To study FQHE systems, one puts N electrons onto the lowest shell ($N \leq 2S+1$, e.g. for $2S = 3(N-1)$ one gets the Laughlin 1/3 state). Under the condition that $\hbar\omega_c \gg e^2/\epsilon a_0$ (Coulomb interaction is small compared to the separation between Landau levels), the problem is limited to the lowest shell states (higher energy excitations are omitted). The number of many-particle states is

$$\frac{(2S+1)!}{N!(2S+1-N)!}. \tag{8.6}$$

The problem is to diagonalize the interaction Hamiltonian (Coulomb interaction)

$$\langle \Psi_0 | H_{\text{int}} | \Psi_0 \rangle \tag{8.7}$$

within subspace of the lowest Landau level, where

$$H_{\text{int}} = \sum_{i<j} \frac{e^2}{\epsilon} |\mathbf{r}_{ij}|^{-1}.$$

The Wigner–Eckart theorem says that energy states are numbered by the total angular momentum (and an additional degeneracy α)

$$\langle \alpha' L' M' | H_{\text{int}} | \alpha L M \rangle = \delta_{LL'} \delta_{MM'} \langle \alpha' L | H_{\text{int}} | \alpha L \rangle. \tag{8.8}$$

In units of $a_0 = \sqrt{\hbar c/eB} = R/\sqrt{S}$ (length) and $e^2/\epsilon a_0$ (energy), the Coulomb interaction takes the simple form $\sum_{i<j} 1/r_{ij}$. The second quantization expression for interaction is:

$$\frac{1}{2} \sum_{m_1=-S}^{S} \sum_{m_2=-S}^{S} \sum_{m_3=-S}^{S} \sum_{m_4=-S}^{S} \left\langle m_1 m_2 \left| \frac{1}{r} \right| m_3 m_4 \right\rangle_S a^+_{m_1} a^+_{m_2} a_{m_3} a_{m_4} \tag{8.9}$$

(operator a^+_m creates a particle in a state $|m\rangle$). The matrix element equals:

$$\left\langle m_1 m_2 \left| \frac{1}{r} \right| m_3 m_4 \right\rangle_S = \sum_{J=0}^{2S} \sum_{M=-J}^{J} \langle S m_1, S m_2 | J M \rangle \langle S m_3, S m_4 | J M \rangle V_J^S R^{-1}, \tag{8.10}$$

where $\langle S m_1, S m_2 | JM \rangle$ are Clebsch–Gordan coefficients (J is the pair angular momentum, $M = -J, -J+1, \ldots, J$) and

$$V_J^S = \frac{\binom{4S-2J}{2S-J}\binom{4S+2J+2}{2S+J+1}}{\binom{4S+2}{2S+1}^2}. \tag{8.11}$$

One can also consider other forms of interaction (e.g. logarithmic) (Wójs and Quinn 2000*a*). Let us consider the case of three electrons onto the angular momentum shell $l = 3$. One has $\binom{7}{3} = 35$ many-particle states. However, when considering total angular momentum states (building so-called Young tableau) one gets five states ($L = 0, 2, 3, 4, 6$)—Table 8.1.

The number of states for higher number of particles are presented in Table 8.2. One can see that the number of states (dimension of a matrix to diagonalize)

Table 8.1 *Young tableau for three particles in angular momentum shell $l = 3$.*

$M = 6$	$M = 5$	$M = 4$	$M = 3$	$M = 2$	$M = 1$	$M = 0$
$\begin{pmatrix} 3 \\ 2 \\ 1 \end{pmatrix}$	$\begin{pmatrix} 3 \\ 2 \\ 0 \end{pmatrix}$	$\begin{pmatrix} 3 \\ 2 \\ -1 \end{pmatrix}$	$\begin{pmatrix} 3 \\ 2 \\ -2 \end{pmatrix}$	$\begin{pmatrix} 3 \\ 2 \\ -3 \end{pmatrix}$	$\begin{pmatrix} 3 \\ 1 \\ -3 \end{pmatrix}$	$\begin{pmatrix} 3 \\ 0 \\ -3 \end{pmatrix}$
		$\begin{pmatrix} 3 \\ 1 \\ 0 \end{pmatrix}$	$\begin{pmatrix} 3 \\ 1 \\ -1 \end{pmatrix}$	$\begin{pmatrix} 3 \\ 1 \\ -1 \end{pmatrix}$	$\begin{pmatrix} 3 \\ 0 \\ -2 \end{pmatrix}$	$\begin{pmatrix} 2 \\ 0 \\ -2 \end{pmatrix}$
			$\begin{pmatrix} 2 \\ 1 \\ 0 \end{pmatrix}$	$\begin{pmatrix} 3 \\ 0 \\ -1 \end{pmatrix}$	$\begin{pmatrix} 2 \\ 0 \\ -1 \end{pmatrix}$	$\begin{pmatrix} 1 \\ 0 \\ -1 \end{pmatrix}$
				$\begin{pmatrix} 2 \\ 1 \\ -1 \end{pmatrix}$	$\begin{pmatrix} 2 \\ 1 \\ -2 \end{pmatrix}$	$\begin{pmatrix} 3 \\ -1 \\ -2 \end{pmatrix}$
						$\begin{pmatrix} 2 \\ 1 \\ -3 \end{pmatrix}$

Table 8.2 *Number of states for the Laughlin 1/3 state* ($2S = 3(N-1)$).

N	$l = S$	$\binom{2S+1}{N}$	N_{L}
3	3	35	5
4	4.5	210	18
5	6	1287	73
6	7.5	8008	338
7	9	50388	1656
8	10.5	319770	8512
9	12	2042975	45207
10	13.5	13123110	246448

grows rapidly with N, in practice numerical calculations for 1/3 state ($2S = 3(N-1)$) are done for $N \leq 12$. Some results for $\nu > 1/3$ are available forhigher N (e.g. $N = 14, 15, 16$). The problem of three or more identical particles all with angular momentum l involves the idea of coefficients of fractional parentage. In the special case of three particles we may write (Rose 1957):

$$|l^3 L^\alpha\rangle = \sum_{L'} F^\alpha_{J'} |l^2(J')lL\rangle,$$

where $|l^2(J')lL\rangle$ is the wave function of the system in which the pair angular momentum $ll(J')$ is defined and α is the additional parameter distinguishing different degenerate states. $F^\alpha_{J'}$ are coefficients of fractional parentage which satisfy the normalization condition $\sum_{J'} F^\alpha_{J'} = 1$. The antisymmetry of the states leads to the matrix equation (Rose 1957):

$$\sum_{J''} F_{J''}(A_{J'',J'} - \delta_{J'',J'}) = 0,$$

where

$$A_{J'',J'} = (-)^{3l-L+1}\sqrt{(2J''+1)(2J'+1)}W(lllL, J''J') \quad \text{and} \quad W(lllL, J''J')$$

are Racah coefficients.[1]

The matrix element of the interaction of the system is given by (de Shalit and Talmi 1963):

$$\left\langle l^3 L^\alpha \middle| \sum_{i<k} V_{ik} \middle| l^3 L^\beta \right\rangle = 3\sum_{L'} F^\alpha_{J'} F^\beta_{J'} V_{ll}(J'), \tag{8.12}$$

where $V_{ll}(J') = V^l_{J'}$ is the pair interaction of particles in $l-l$ coupling in the state of the pair angular momentum J' (8.11).

Let us also comment on the system of three particles, two of them with angular momentum l, the third with angular momentum l'. The wave function of the system in the state of the total angular momentum L can be written using the Racah decomposition (Rose 1957):

$$|l^2(J_1)l'L\rangle = \sum_{J'} R_{J_1,J'} |ll'(J')lL\rangle. \tag{8.13}$$

[1]Racah coefficients can be obtained in a straightforward manner from Clebsh–Gordan coefficients (3-j symbols). We have the following relation between W coefficients and 6-j symbols:

$$W(j_1 j_2 l_2 l_1; j_3 l_3) = (-1)^{j_1+j_2+l_1+l_2} \begin{pmatrix} j_1 & j_2 & j_3 \\ l_1 & l_2 & l_3 \end{pmatrix}.$$

Here $|l^2(J_1)l'L\rangle$ denotes the state in which two particles of angularmomentum l are coupled to give a resulting two particle angular momentum J_1; J_1 is then coupled with l' to give a total angular momentum L. In Eqn (8.13)

$$R_{J_1,J'} = \sqrt{(2J_1+1)(2J'+1)}W(llLl', J_1J'),$$

where W are Racah coefficients.[2]

The matrix element of the interaction energy is

$$\left\langle l^2(J_1)l'L \middle| \sum_{i<k} V_{ik} \middle| l^2(J_2)l'L \right\rangle.$$

We separate the interaction in the same shell (ll) and the interactions between shells ll' (de Shalit and Talmi 1963); then this matrix element becomes

$$= V_{ll}(J_1)\delta_{J_1,J_2} + 2\sum_{J'} R_{J_1,J'}R_{J_2,J'}V_{ll'}(J'). \tag{8.14}$$

In the case of degenerate states the interaction energies of states are eigenvalues of the interaction matrix. Eqns (8.12) and (8.14) allow us to obtain the energy of three particle states in terms of two-particle interactions. The exact diagonalization for higher number of particles goes in a similar way (Wójs and Quinn 2000*a*)—see Chapter 9. The results of exact diagonalization exhibit the structure of Laughlin (Fano *et al.* 1986; He *et al.* 1992) (Fig. 8.1) and Jain states (Dev and Jain 1992; He *et al.* 1994). Considering the trial wave functions it is interesting to note that for the three electron case the Laughlin wave function is the exact ground state for the 1/3 state (Haldane 1983) (in contrast to the disc geometry (Laughlin 1983*b*)).

8.2 Composite fermion transformation

The composite fermion transformation consists in replacing the real magnetic field by the effective field in the manner:

$$2S^* = 2S - 2p(N-1) \tag{8.15}$$

($2p$—strength of the Chern–Simons field). The energy spectra in Figs 8.1 and 8.2 can be readily interpreted in terms of composite fermions (Chen and Quinn 1994) (the limitation of N makes it difficult to observe many Jain states, and also makes it difficult to exclude some of the non-Jain states, e.g. 4/11 (Sitko *et al.* 1997*b*)). The interpretation goes as follows: the system of N electrons is interpreted in

[2] There are some symmetry relations for the Racah coefficients, and 'sum rules' (de Shalit and Talmi 1963), for example,

$$\sum_e R_{\mathrm{fe}}R_{\mathrm{ge}} = \delta_{\mathrm{fg}}, \quad \sum_e R_{\mathrm{ef}}R_{\mathrm{eg}} = \delta_{\mathrm{fg}}.$$

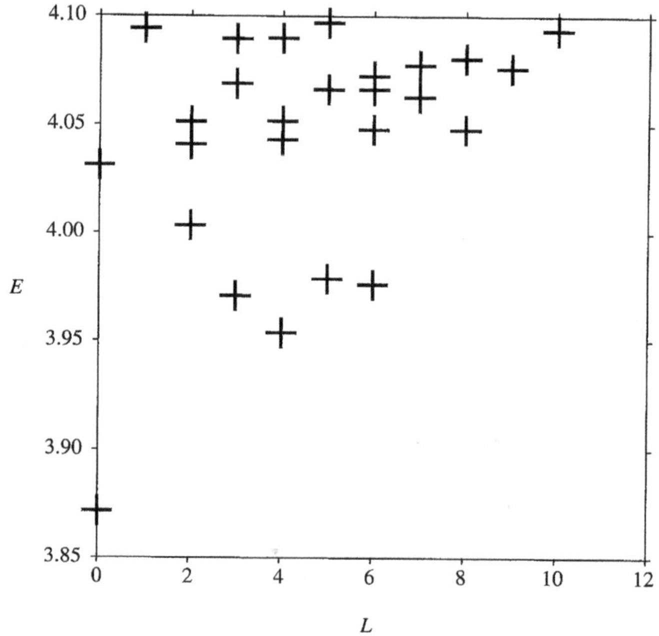

FIG. 8.1. Energy spectrum for $N = 6$, $2S = 15$ (the Laughlin 1/3 state).

terms of N (composite) fermions in the field $2S^*$. For the Laughlin states one always gets one completely filled effective shell:

$$2S^* = N - 1. \tag{8.16}$$

For example, for $N = 6$ and $2S = 15$ (Fig. 8.1), the lowest energy state is the $L = 0$ state (as it would be if the lowest shell would be completely filled—as it is the case in the field $2S^* = 5$). The higher energy states can also be interpreted using standard methods of the shell theory (de Shalit and Talmi 1963).[3] We can see a band of states with $L = 2, 3, 4, 5, 6$—exactly as predicted by QE($l = S^* + 1 = 3.5$) – QH($l = S^* = 2.5$) (QE, quasi electrons; QH, quasiholes) ($L = 1, 2, 3, 4, 5, 6$) excitations (note that $L = 1$ QE–QH state is missing).[4] Similarly, in Fig. 8.2, $2S = 16$, (one flux quantum is added to the system—one QH created) $2S^* = 6$ and there is an empty place in the effective shell—a single

[3]The main idea is that the shell with $2l + 1 - n$ particles can be treated in terms of the conjugated configuration of l^n quasiholes (QH). The allowed angular momenta of these two configurations are the same. The idea of QH can be extended to the configuration of particles in two shells. For example, the $l^{2l}l'$ configuration can be considered as a particle–hole pair. One can treat the particle–hole interaction $V_{l^{2l}l'}$ as the interaction between two particles ll', $V_{ll'}$ (de Shalit and Talmi 1963).

[4]The missing of $L = 1$ QE $-$ QH state is predicted by Dev and Jain within the composite fermion trial wave function approach (Dev and Jain 1992) (due to the lack of overlap of projected wave function).

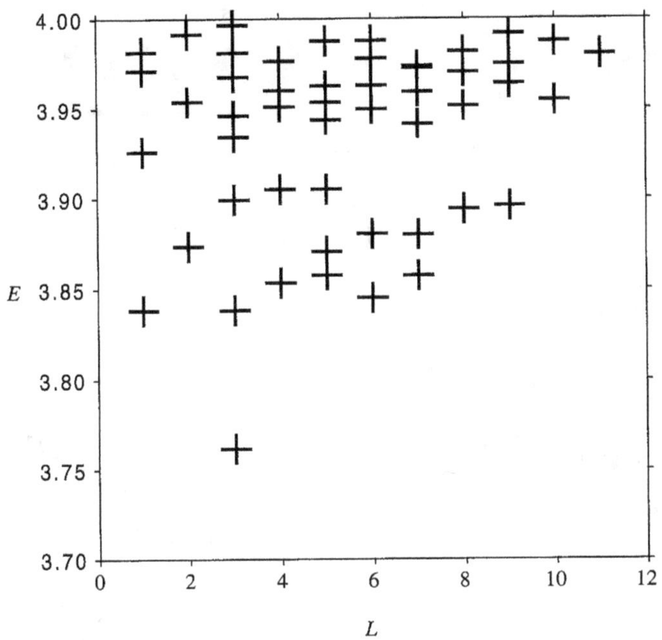

FIG. 8.2. Energy spectrum for $N = 6$, $2S = 16$.

QH (of Laughlin type)—a composite fermion excitation. This QH (almost filled shell) has angular momentum $l = 3$ and we can see in Fig. 8.2 that this is in fact the lowest energy state. Similar interpretation holds for other values of $2S$ and also for higher energy states (Chen and Quinn 1994; Sitko *et al.* 1996). For Jain states $n/(2pn+1)$, n effective shells are completely filled (e.g. the 2/5 state). In this way the spectra of QE–QH, 1QE, 1QH, 2QE–QH, 2QH–QE, 2QE, 2QH, 3QE, 3QH, N_{QE}, N_{QH} can be interpreted (Chen and Quinn 1994; Sitko *et al.* 1996). The attempts to quantitatively interpret the spectra involve the introduction of quasiparticle interaction. Here we present the way to get phenomenological interaction (Yi *et al.* 1996) (in contrast to more detailed analysis of Beran and Morf (1991))—in analogy to Fermi liquid interaction (Pines and Nozieres 1966):

$$E_{2\mathrm{QE}} = 2\epsilon_{\mathrm{QE}} + V_{\mathrm{QE\text{-}QE}}. \tag{8.17}$$

The data can be extracted from numerical results for 2QE, 2QH, and then tested for N_{QP} systems (single quasiparticle (QP) energy is extracted from data for single QE and single QH—when compared with the Laughlin ground state (Yi *et al.* 1996)). Such a procedure was introduced in Sitko *et al.* (1996) and the results appeared to be in a good qualitative agreement—see Fig. 8.3 for comparison of the 'Fermi liquid' shell model (circles) and numerical results.

In a similar way the energy of the 2/5 Jain state can be calculated starting from the Laughlin 1/3 state (thus creating so many quasielectrons to fill the

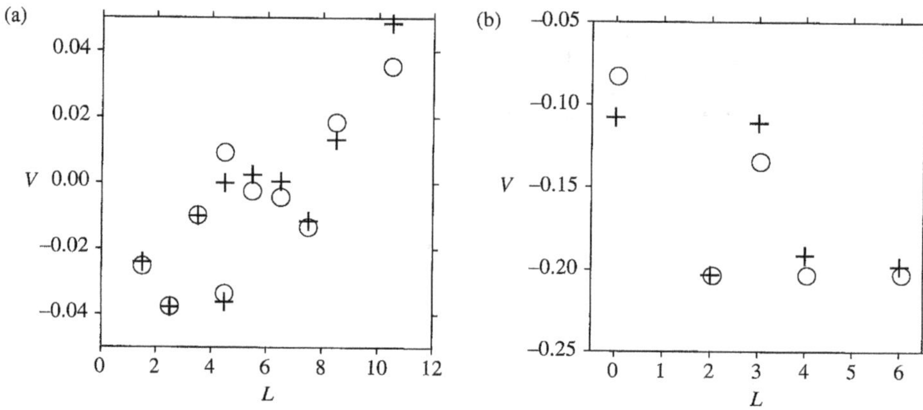

FIG. 8.3. Energy spectrum for (a) $3QH$, $N = 7$ and (b) $3QE$, $N = 8$.

Table 8.3 *Results of the 'Fermi liquid' shell model approach for the 2/5 state (energy in units e^2/a_0, where a_0 is the magnetic length for the Laughlin 1/3 state). The interaction energy of a closed shell is $V(l^{2l+1}\ L = 0) = \sum_{L'}(2L' + 1)V_{ll}(L')$.*

N	1/3	2/5	
	Exact	Exact	Predicted
6	3.872	4.002	4.004
8	6.363	6.522	6.535

next effective shell). Comparison of the exact ground-state energy of the state of 2/5 with the shell model results is presented in Table 8.3. In analogy to the half-filled state (Halperin *et al.* 1993), Rezayi and Read (1994) suggested to look at the state given by $2S = 2(N - 1)$ (zero effective field $2S^* = 0$). The results for $N = 9$ are given in Fig. 8.4. Let us consider the case of $N = 9$; then for $2S = 16$ one has $2S^* = 0$, so that in terms of composite fermions there is no magnetic field, the lowest effective shell is for $l = 0$ (with one particle). Nine particles fill up three shells ($l = 0$, $l = 1$, $l = 2$). The lowest energy state should be $L = 0$ state, as we can see in Fig. 8.4. First excited states would be the particle ($l = 3$) – hole ($l = 2$) excitation with $L = 1, 2, 3, 4, 5$. We see those states in Fig. 8.4 (except $L = 1$ state—again as for the QE–QH excitation in the Laughlin state (Dev and Jain 1992)). Similar interpretation holds for $N = 10$ and $N = 11$ (Rezayi and Read 1994).

We should mention that there are some difficulties in the interpretation of the $2S = 2(N - 1)$ state as the half-filled state (Jain and Kamilla 1997; Morf 1998) (as refered to the composite fermion compressible state introduced by Halperin,

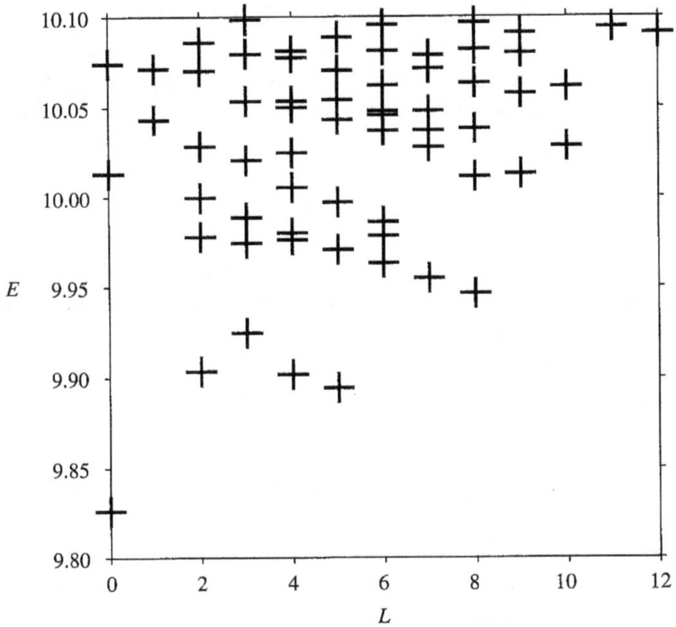

FIG. 8.4. Energy spectrum for $N = 9$, $2S = 16$ (the half-filled state).

Lee, and Read (HLR)), only the $N \longrightarrow \infty$ limit corresponds to the HLR Fermi liquid.

8.3 Hierarchy

Let us now comment on the case when $N > 2S + 1$ and $N - (2S + 1) < 2S + 1 + 2$—then the system is effectively the system of $N - (2S + 1)$ quasielectrons, such a description is valid when the separation between Landau levels $\sim \hbar\omega_c$ is large enough (compared to the interaction energy e^2/a_0) and only two-body interactions are present (closed shells can be ignored—they give nothing but a constant shift in energy). The problem is analogous to the starting one of electrons partially filling lowest Landau level. For higher levels, the degeneracy of the nth Landau level is $2S + 1 + 2n$ (the interaction in higher shells changes, however; Wójs and Quinn (2000*a*)).

Within the composite fermion approach one deals with effective shells. If ignoring of the filled effective shells is still valid, then the only important effective shell is the partially filled one (for the lowest energy sector, which is the part of the energy states separated by the well-defined energy gap—this is what we see in Figs 8.1–8.4). If, again, a partially filled effective shell can be treated as an electron shell were treated, then the composite fermion transformation should predict condensed states of quasiparticles. The open question is whether this is the case (Haldane 1983; Sitko *et al.* 1997*b*; Wójs and Quinn 2000*a*) (recently, indications for several hierarchical states were found in the experiment (Pan *et al.* 2003)—see Fig. 9.6). Applying composite fermion transformation to electrons, the filling fraction can be written as $\nu_0^{-1} = 2p_0 + (1/(n_1 + \nu_1))$, where ν_1 is the filling of the partially filled effective shell. Considering quasiparticles (composite fermion excitations) we write:

$$2S^*_{\mathrm{QE}} = 2S_{\mathrm{QE}} - 2p_{\mathrm{QE}}(n_{\mathrm{QE}} - 1)$$

and ask whether $2S^*_{\mathrm{QE}}$ would describe the condensed states of quasielectrons. If this shell can be treated in the same way as the partially filled electron shell, then $\nu_1^{-1} = 2p_1 + (1/(n_2 + \nu_2))$ should predict fillings (if $\nu_2 = 0$)—analogs of Laughlin and Jain states. Repeating this procedure until at final step $\nu_m = 0$, one gets the composite fermion hierarchy fraction:

$$n_0 = \frac{1}{2p_0 + 1/(n_1 + (1/2p_1 + \cdots))}. \tag{8.18}$$

It can be shown that: Eqn (8.18) is the Haldane hierarchy (Haldane 1983) when all $n_l = 1$,[5] Jain states are found when $\nu_1 = 0$, the Haldane hierarchy is equivalent to Eqn (8.18) (Sitko *et al.* 1997*b*).[6]

Let us consider a spherical system containing n_l^{QE} quasielectrons at the lth hierarchical level. If those particles occupy one filled Landau shell and there are extra particles in the partially filled second Landau level, one finds for the lowest angular momentum shell $2S_{l+1} - 1 = n_l^{\mathrm{QE}} - n_{l+1}^{\mathrm{QE}}$ (Haldane 1983; Chen and Quinn 1994). A condensed state (of Laughlin type) of these quasiparticles is obtained when they fill one shell at the next step of the hierarchy ($n_{l+2}^{\mathrm{QE}} = 0$); then we have additionally $2S_{l+1} = (2p_{l+1} + 1)(n_{l+1}^{\mathrm{QE}} - 1)$. Combining these two relations, we get

$$n_l^{\mathrm{QE}} = (2p_{l+1} + 2)\left(n_{l+1}^{\mathrm{QE}} - 1\right) \tag{8.20}$$

in agreement with Haldane (1983) (if quasiholes are produced then: $n_l^{\mathrm{QE}} = 2p_{l+1}(n_{l+1}^{\mathrm{QH}} - 1)$). Thus, we identify Haldane even numbers as $2p_{l+1} + 2$ for quasielectrons ($2p_{l+1}$ for quasiholes), the odd number is given by $2p_0 + 1$.

[5]Proof: since

$$\nu_l^{-1} = 2p_l + 1 - (1 + \nu_{l+1}^{-1})^{-1}$$

one gets the form of the Haldane hierarchy fraction (Haldane 1983).

[6]Proof: put all $p_l = 0$, then (if $\nu_n = 0$)

$$\frac{1}{1 - 1/(2 - 1/(2 - \cdots - 1/2))} = n$$

or with $p_0 \neq 0$

$$\frac{1}{2p_0 + 1 - 1/(2 - 1/(2 - \cdots - 1/2))} = \frac{n}{2p_0 n + 1}. \tag{8.19}$$

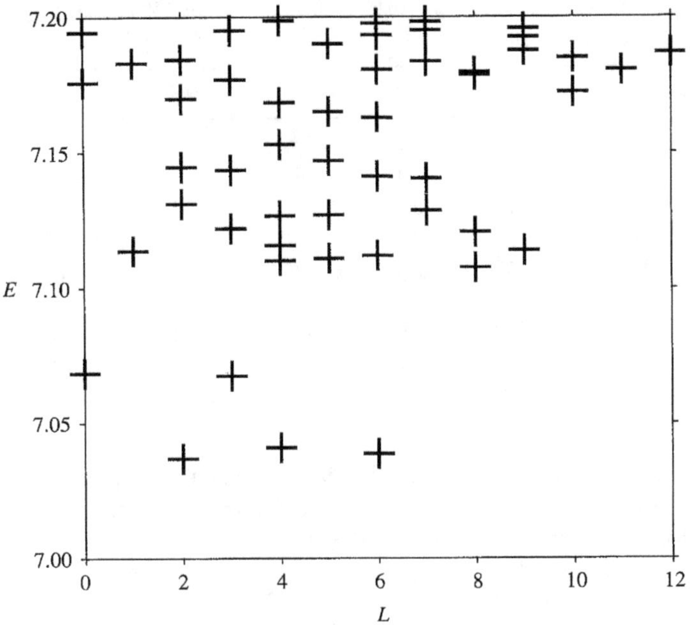

FIG. 8.5. Energy spectrum for $N = 8$, $2S = 18$ (the $\nu = 4/11$ state).

The construction of the Haldane hierarchy for quasielectrons without applying the composite fermion transformation to them, that is, $p_{l+1} = 0$ (redefinition of the number of quasielectrons), leads to the Jain principal fraction (8.19), and the picture becomes identical to the original Jain construction.

An interesting example is the 4/11 state. The 4/11 state is the 1/3 state of quasielectrons of the Laughlin 1/3 state (one may compare with the 2/7 Jain state—the 1/3 state of quasiholes). Numerical diagonalization results for small systems of electrons are presented in Fig. 8.5. They show no indication of condensed state (it is interesting that quasiholes at 1/3 filling—2/7 Jain state—form a condensed state but quasielectrons do not (Sitko *et al.* 1997*b*)). However, the 4/11 state was recently observed (and indications for 5/13, 4/13, 5/17, 6/17, 7/11 states were found—see Fig. 9.6) in the experiment (Pan *et al.* 2003).

Despite the fact that hierarchical states are only hypothetical (several hierarchy fractions were recently observed) the method described above can be used to define the filling fraction parameter. Putting together the above ideas, one can define the way of getting the filling fraction:

$$\frac{1}{\nu} = 2p + \frac{\alpha}{n + \nu_{\mathrm{QE}}}, \tag{8.21}$$

where α reflects the sign of the effective field, n, the number of filled effective

Table 8.4 *The filling fraction for $N = 12$ electrons on a Haldane sphere.*

$2S$	ν	$2S$	ν	$2S$	ν
11	1	19	15/23	27	13/33
12	—	20	—	28	—
13	12/13	21	3/5	29	4/11
14	6/7	22	—	30	6/17
15	4/5	23	3/7	31	12/35
16	—	24	—	32	—
17	13/19	25	15/37	33	1/3
18	2/3	26	2/5	34	—

shells, $\nu_{\rm QE}$, the filling fraction for the partially filled shell of quasielectrons. When one repeats Eqn (8.21) until at mth step and $\nu_{\rm QE}^m = 0$, the odd denominator hierarchy fractions are obtained or $\alpha^m = 0$—even denominator hierarchy fractions, generalized half-filled states (Rezayi and Read 1994) (the experimental evidence for first even-denominator hierarchical states 3/8 and 3/10 were recently found in the experiment (Pan *et al.* 2003)—see Fig. 9.6). The method omits only those cases where there is one particle left (quasielectron). The results for $N = 12$ are given in the Table 8.4. The filling for particle–hole states always adds to one, except for three cases (Sitko *et al.* 1997*a*; Morf 1998)

$$N_e = N_h(\nu_e = \nu_h < 1/2)$$
$$N_e = N_h \pm 1(+\nu_e = 1/2, \nu_h < 1/2, N_h < N_e)$$

The structure of the filling fraction can be seen calculating the so-called finite size corrections (D'Ambrumanil and Morf 1989; Morf 1998) which we define as:

$$\frac{1}{\nu} - \frac{2S}{N-1}$$

for a constant number of particles $N = 12$, within range $1 \geq \nu \geq 1/5$ is shown in Fig. 8.6. Let us comment on the shape of the curves in Fig. 8.6. First, what are easily noticed are the discontinuities at 'half-filled' ($\nu = 1/2p$) states. This can be easily explained when looking at the Jain fractions (Sitko *et al.* 1997*a*)

$$\frac{2S}{N-1} = \frac{2pn+1}{n} - \frac{n^2-1}{n(N-1)}, \quad \text{for } 2S^* > 0,$$

and

$$\frac{2S}{N-1} = \frac{2pn-1}{n} + \frac{n^2-1}{n(N-1)}, \quad \text{for } 2S^* < 0.$$

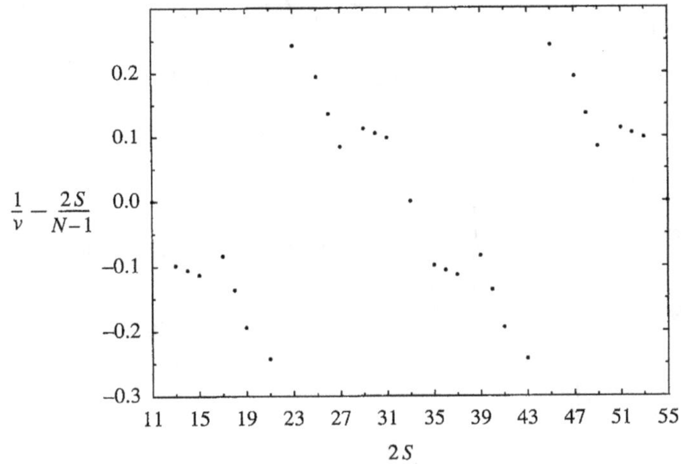

FIG. 8.6. Finite size corrections for $N = 12$ electrons on a sphere.

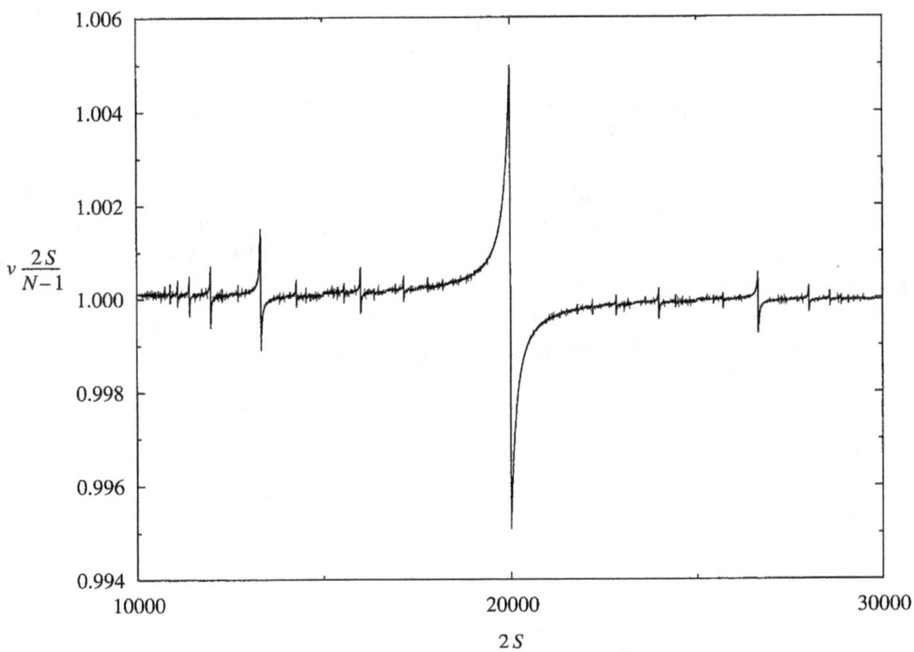

FIG. 8.7. Finite size corrections for $N = 10001$ electrons on a sphere, $1/3 \leq \nu \leq 1$.

The corrections are of opposite sign. The result is supported by other hierarchy fractions. All 'half-filled' states (even denominator fractions) are seen as discontinuities. The 'curve' repeats itself when $p = 2$. We also plot the results for

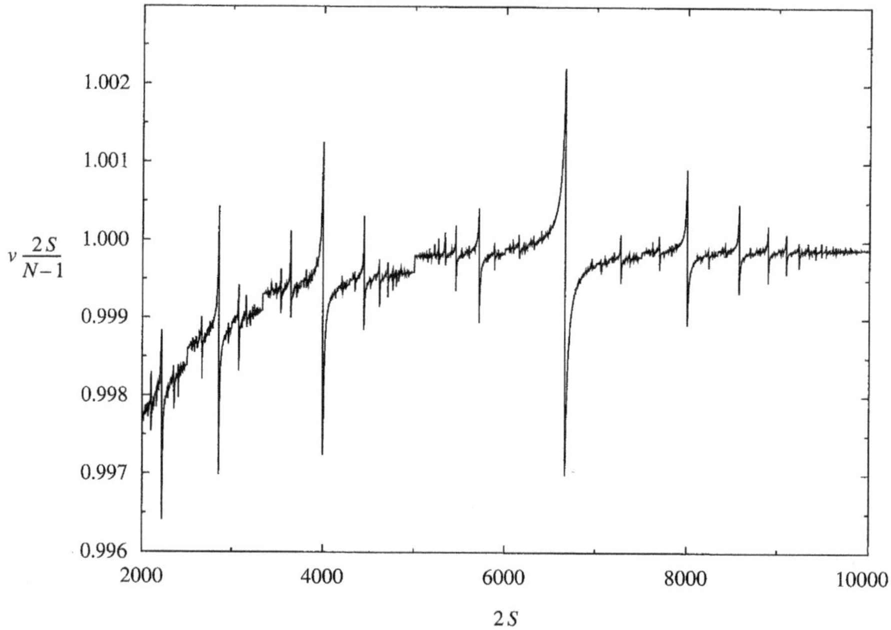

FIG. 8.8. The filling fraction for $N = 10001$, $1 \leq \nu$.

$N = 10001$ in Figs 8.7 and 8.8 (fillings $\nu > 1$). In Fig. 8.8, the integer fractions are seen as small jumps. Such jumps also appear at Laughlin and Jain states (Fig. 8.7). When resolution (scale, number of particles, $2S$) increases, all odd denominator fractions lead to such 'jumps'. At any scale, the picture remains much the same—it has a fractal structure. The filling fraction exhibits in itself the structure of the FQHE (Sitko *et al.* 1997*b*, 1998).

9

PSEUDOPOTENTIAL APPROACH TO THE FRACTIONAL QUANTUM HALL STATES

9.1 Problems with justification of the composite fermion picture

In the mean field composite fermion picture, a two-dimensional electron gas at the density ρ and in a strong external magnetic field B is converted into a system of composite fermions at the same density ρ, but at a reduced, effective magnetic field, B^*. The resulting (average) effective magnetic field seen by a composite fermion is $B^* = B - 2p\phi_0\rho$. Because $B^*\nu^* = B\nu = \rho\phi_0$, the relation between the electron and composite fermion filling factors (defined as the number of particles, N, divided by their Landau level degeneracy) is

$$(\nu^*)^{-1} = \nu^{-1} - 2p. \tag{9.1}$$

Since the low band of energy levels of the original (interacting) two-dimensional electron gas has a structure similar to that of the non-interacting composite fermions in a uniform effective field B^*, it was proposed (Jain 1989) that the Coulomb charge–charge and Chern–Simons (CS) charge–flux interactions beyond the mean field largely cancel one another, and the original strongly interacting system of electrons is converted into one of weakly interacting composite fermions. Consequently, the fractional quantum Hall effect (FQHE) of electrons was interpreted as the integral quantum Hall effect (IQHE) of composite fermions.

Although the mean field composite fermion picture correctly predicts the structure of low energy spectra of fractional quantum Hall systems, its energy scale (the composite fermion cyclotron energy $\hbar\omega_c^*$) is totally irrelevant. Moreover, since the characteristic energies of CS ($\hbar\omega_c^* \propto B$) and Coulomb ($e^2/\lambda \propto \sqrt{B}$, where λ is the magnetic length) interactions between fluctuations beyond mean field scale differently with the magnetic field, the reason for the success of the composite ferminon model cannot be found in originally suggested cancellation between those interactions. Since the mean field composite fermion picture is commonly used to interpret various numerical and experimental results, it is very important to understand why and under what conditions it is correct.

In this chapter, we use the pseudopotential formalism (Haldane and Rezayi 1988; Wójs and Quinn 1998, 1999 2000*c*) to describe the fractional quantum Hall states. We show that the form of the pseudopotential $V(L')$, defined as the pair interaction energy V expressed as a function of the pair angular momentum L', rather than the interaction potential $V(r)$, is responsible for the incompressibility of fractional quantum Hall states. The idea of fractional parentage (de Shalit and Talmi 1963; Cowan 1981) is used to characterize many-body states by the ability of electrons to avoid pair states with largest repulsion. The strict condition on the form of pseudopotential $V(L')$ necessary for the occurrence of fractional quantum Hall states is given, which defines the class of short-range pseudopotentials to which mean field composite fermion picture can be applied. In a number of examples, we explain the success or failure of the mean field composite fermion predictions for the system of electrons in the lowest and excited Landau levels, for various Laughlin quasiparticles (QPs) in the hierarchy picture of fractional quantum Hall states (Haldane 1983; Sitko *et al.* 1997*b*), and for charged excitons in a two-dimensional electron–hole plasma (Wójs *et al.* 1998, 1999*a*, *b*).

9.2 Numerical studies on the Haldane sphere

Due to the macroscopic Landau level degeneracy, the electron–electron interaction in the fractional quantum Hall states cannot be treated perturbatively. Therefore, the exact (numerical) diagonalization techniques have been commonly used to study these systems. In order to model an infinite two-dimensional electron gas by a finite (small) system that can be handled numerically, it is very convenient to confine N electrons to a surface of a (Haldane) sphere (Haldane 1983) of radius R. In this model, the normal magnetic field B is produced by a fictitious Dirac magnetic monopole placed in the centre of the sphere. The monopole strength $2S$ is quantized as reired by the Dirac condition and, when expressed in the units of the flux quantum $\phi_0 = hc/e$, it can only take on integer values, $2S = 0, 1, 2, \ldots$ In these units, the total flux through the surface of the sphere is $4\pi BR^2 = 2S\phi_0$, and the magnetic length is related to the radius of the sphere by $R^2 = S\lambda^2$.

The obvious advantages of spherical geometry are the absence of an edge and preserving full two-dimensional symmetry of a two-dimensional electron gas. The pair of good quantum numbers on the sphere are the total angular momentum L and its projection M. They can be related to the pair of good quantum numbers on the plane, the total and centre-of-mass angular momenta. The numerical experiments in the spherical geometry have shown that even relatively small systems that can be solved exactly on a small computer behave in many ways like an infinite two-dimensional electron gas, and a number of parameters of a two-dimensional electron gas (e.g. characteristic excitation energies) can be obtained from such small-scale calculations.

The single-particle states on a Haldane sphere (monopole harmonics) are labelled by angular momentum l and its projection m (Wu and Yang 1976). The energies, $\varepsilon_l = \hbar\omega_c[l(l+1) - S^2]/2S$, fall into degenerate shells and the nth shell ($n = l - |S| = 0, 1, \ldots$) corresponds to the nth Landau level. For the fractional quantum Hall states at filling factor $\nu < 1$, only the lowest, spin polarized Landau level need be considered.

The object of numerical studies is to diagonalize the electron–electron interaction hamiltonian H in the space of degenerate antisymmetric N electron states of a given (lowest) Landau level. Although matrix H is easily block diagonalized into blocks with specified M, the exact diagonalization becomes difficult (matrix dimension over 10^6) for $N > 10$ and $2S > 27$ ($\nu = 1/3$) (Wójs and Quinn 1998, 1999, 2000c). Typical results for 10 electrons at filling factors near $\nu = 1/3$ are presented in Fig. 9.1. Energy E, plotted as a function of L in the magnetic units, includes shift $-(Ne)^2/2R$ due to charge compensating

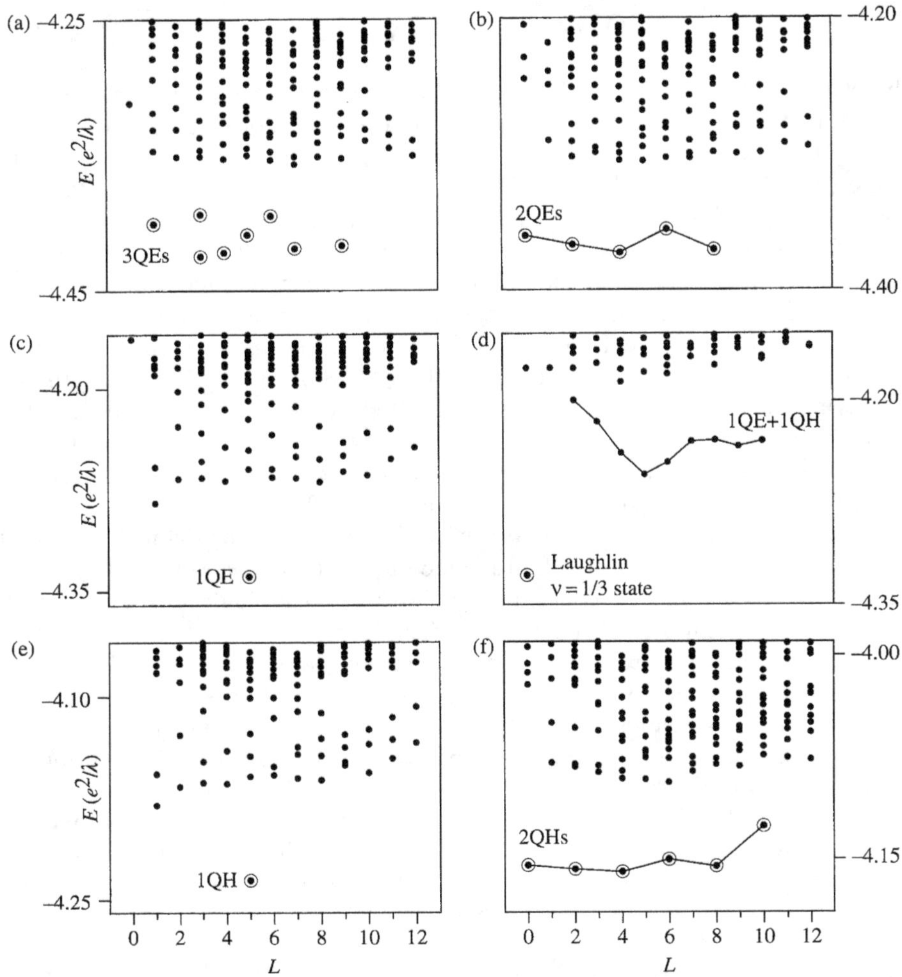

FIG. 9.1. Energy spectra of 10 electrons in the lowest Landau level at the monopole strength $2S$ between 24 and 29. Open circles mark lowest energy bands with fewest composite fermion QPs. (a) $2S = 24$, (b) $2S = 25$, (c) $2S = 26$, (d) $2S = 27$, (e) $2S = 28$, (f) $2S = 29$.

background. There is always one or more L multiplets (marked with open circles) forming a low energy band separated from the continuum by a gap. If the lowest band consists of a single $L = 0$ ground state (Fig. 9.1(d)), it is expected to be incompressible in the thermodynamic limit (for $N \to \infty$ at the same ν) and an infinite two-dimensional electron gas at this filling factor is expected to exhibit the fractional quantum Hall effect.

The mean field composite fermion interpretation of the spectra in Fig. 9.1 is the following. The effective magnetic monopole strength seen by composite fermions is (Jain 1989; Wójs and Quinn 1998, 1999, 2000*c*)

$$2S^* = 2S - 2p(N-1), \tag{9.2}$$

where $N-1$ instead of N reflects the fact that each composite fermion only couples to the magnetic flux on other composite fermions but not to its own flux, and the angular momenta of lowest composite fermion shells (composite fermion Landau levels) are $l_n^* = |S^*| + n$ (Chen and Quinn 1994). At $2S = 27$, $l_0^* = 9/2$ and 10 composite fermions fill completely the lowest composite fermion shell ($L = 0$ and $\nu^* = 1$). The excitations of the $\nu^* = 1$ composite fermion ground state involve an excitation of at least one composite fermion to a higher composite fermion Landau level, and thus (if the composite fermion–composite fermion interaction is weak on the scale of $\hbar\omega_c^*$), the $\nu^* = 1$ ground state is incompressible and so is Laughlin (Laughlin 1983*b*) $\nu = 1/3$ ground state of underlying electrons. The lowest lying excited states contain a pair of QPs: a quasihole (QH) with $l_{\text{QH}} = l_0^* = 9/2$ in the lowest composite fermion Landau level and a quasielectron (QE) with $l_{\text{QE}} = l_1^* = 11/2$ in the first excited one. The allowed angular momenta of such pair are $L = 1, 2, \ldots, 10$. The $L = 1$ state usually has high energy and the states with $L \geq 2$ form a well-defined band with a magnetoroton minimum at a finite value of L. The lowest composite fermion states at $2S = 26$ and 28 contain a single QE and a single QH, respectively (in the $\nu^* = 1$ composite fermion state, i.e. the $\nu = 1/3$ electron state), both with $l_{\text{QP}} = 5$, and the excited states will contain additional QE–QH pairs. At $2S = 25$ and 29, the lowest bands correspond to a pair of QPs, and the values of energy within those bands define the QP–QP interaction pseudopotential V_{QP}. At $2S = 25$, there are two QEs each with $l_{\text{QE}} = 9/2$ and the allowed angular momenta (of two identical fermions) are $L = 0, 2, 4, 6$, and 8, while at $2S = 29$ there are two QHs each with $l_{\text{QH}} = 11/2$ and $L = 0, 2, 4, 6, 8$, and 10. Finally, at $2S = 24$, the lowest band contains three QEs each with $l_{\text{QE}} = 4$ in the Laughlin $\nu = 1/3$ state (in the Fermi liquid picture, interacting with one another through V_{QE} (Sitko *et al.* 1996)) and $L = 1, 3^2, 4, 5, 6, 7$, and 9.

9.3 Pseudopotential approach

The two-body interaction hamiltonian H can be always expressed as

$$\hat{H} = \sum_{i<j} \sum_{L'} V(L') \hat{\mathcal{P}}_{ij}(L'), \tag{9.3}$$

where $V(L')$ is the interaction pseudopotential (Haldane and Rezayi 1988) and $\hat{\mathcal{P}}_{ij}(L')$ is the projection operator onto the subspace with angular momentum of pair ij equal to L'. For electrons confined to a Landau level, L' measures the average squared distance d^2 (Wójs and Quinn 1998, 1999, 2000*c*),

$$\frac{\hat{d}^2}{R^2} = 2 + \frac{S^2}{l(l+1)}\left(2 - \frac{\hat{L}'^2}{l(l+1)}\right), \tag{9.4}$$

and larger L' corresponds to a smaller separation. Remarkably, the severe limitation of the available Hilbert space imposed by the confinement of electrons to one (lowest) Landau level, results in the fact that the interaction potential $V(r)$ enters hamiltonian H only through a small number of pseudopotential parameters $V(2l-\mathcal{R})$, where $\mathcal{R}$, relative pair angular momentum, is (for indistinguishable fermions) an odd integer.

In Fig. 9.2 we compare Coulomb pseudopotentials $V(L')$ calculated for a pair of electrons on the Haldane sphere each with $l = 5$, $15/2$, 10, and $25/2$, in the lowest and in the first excited Landau level. For the reason that will become clear later, $V(L')$ is plotted as a function of squared angular momentum, $L'(L'+1)$. All pseudopotentials in Fig. 9.2 increase with increasing L'. If $V(L')$ increased very quickly with increasing L' (we define ideal short-range repulsion as: $dV_{\rm SR}/dL' \gg 0$ and $d^2V_{\rm SR}/dL'^2 \gg 0$), the low lying many-body states would be the ones maximally avoiding pair states with largest L' (Haldane and Rezayi 1988; Wójs and Quinn 1998, 1999, 2000*c*). At filling factor $\nu = 1/m$ (m is odd) the many-body Hilbert space contains exactly one multiplet in which all pairs completely avoid states with $L' > 2l - m$. This multiplet is the $L = 0$

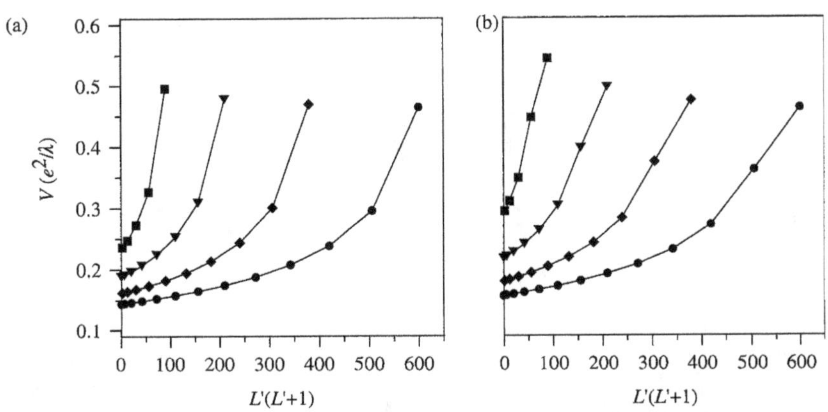

FIG. 9.2. Pseudopotentials V of the Coulomb interaction in the (a) lowest ($n = 0$) and (b) first excited Landau level ($n = 1$) as a function of squared pair angular momentum $L'(L'+1)$. Squares ($l = 5$), triangles ($l = 15/2$), diamonds ($l = 10$), and circles ($l = 25/2$) mark data for different $S = l + n$.

incompressible Laughlin state (Laughlin 1983*b*) and it is an exact ground state of V_{SR}.

The ability of electrons in a given many-body state to avoid strongly repulsive pair states can be conveniently described using the idea of fractional parentage (de Shalit and Talmi 1963; Cowan 1981; Wójs and Quinn 1998, 1999, 2000*c*). An antisymmetric state $|l^N, L\alpha\rangle$ of N electrons each with angular momentum l that are combined to give total angular momentum L can be written as:

$$|l^N, L\alpha\rangle = \sum_{L'} \sum_{L''\alpha''} G_{L\alpha,L''\alpha''}(L')\, |l^2, L'; l^{N-2}, L''\alpha''; L\rangle . \tag{9.5}$$

Here, $|l^2, L'; l^{N-2}, L''\alpha''; L\rangle$ denote product states in which $l_1 = l_2 = l$ are added to obtain L', $l_3 = l_4 = \cdots = l_N = l$ are added to obtain L'' (different L'' multiplets are distinguished by an additional label α''), and finally L' is added to L'' to obtain L. The idea behind the above expansion of the state $|l^N, L\alpha\rangle$ in terms of states $|l^2, L'; l^{N-2}, L''\alpha''; L\rangle$ is that the former state is totally antisymmetric, and the latter states are antisymmetric under interchange of particles 1 and 2, and under interchange of any pair of particles 3, 4, ..., N. This property, together with the fact that (in the Hilbert space of states that are totally symmetric) every two-body (interaction) operator $\sum_{i<j} V_{ij}$ can be replaced by $\frac{1}{2}N(N-1)V_{12}$, allows convenient expression of the many-body matrix elements through the factors $G_{L\alpha,L''\alpha''}(L')$, called the coefficient of fractional grandparentage (composite fermion GP). The two-body interaction matrix element expressed through composite fermion GPs is

$$\langle l^N, L\alpha|\, V\, |l^N, L\beta\rangle = \frac{N(N-1)}{2} \sum_{L'} \sum_{L''\alpha''} G_{L\alpha,L''\alpha''}(L') G_{L\beta,L''\alpha''}(L')\, V(L'), \tag{9.6}$$

and the expectation value of energy is

$$E_\alpha(L) = \frac{N(N-1)}{2} \sum_{L'} \mathcal{G}_{L\alpha}(L')\, V(L'), \tag{9.7}$$

where the coefficient

$$\mathcal{G}_{L\alpha}(L') = \sum_{L''\alpha''} |G_{L\alpha,L''\alpha''}(L')|^2 \tag{9.8}$$

gives the probability that any pair ij is in the state with angular momentum L'.

9.4 Energy spectra of short-range pseudopotentials

The very good description of actual ground states of a two-dimensional electron gas at fillings $\nu = 1/m$ by the Laughlin wave function (overlaps typically larger that 0.99) and the success of the mean field composite fermion picture at $\nu < 1$ both rely on the fact that pseudopotential of Coulomb repulsion in the lowest Landau level falls into the same class of short-range pseudopotentials as $V_{\rm SR}$. Due to a huge difference between all parameters $V_{\rm SR}(L')$, the corresponding many-body Hamiltonian has the following hidden symmetry: the Hilbert space $\mathcal{H}$ contains eigensubspaces $\mathcal{H}_p$ of states with $\mathcal{G}(L') = 0$ for $L' > 2(l-p)$, that is, with $L' < 2(l-p)$. Hence, $\mathcal{H}$ splits into subspaces $\tilde{\mathcal{H}}_p = \mathcal{H}_p \setminus \mathcal{H}_{p+1}$, containing states that do not have grandparentage from $L' > 2(l-p)$, but have some grandparentage from $L' = 2(l-p)-1$,

$$\mathcal{H} = \tilde{\mathcal{H}}_0 \oplus \tilde{\mathcal{H}}_1 \oplus \tilde{\mathcal{H}}_2 \oplus \cdots \tag{9.9}$$

The subspace $\tilde{\mathcal{H}}_p$ is not empty (some states with $L' < 2(l-p)$ can be constructed) at filling factors $\nu \le (2p+1)^{-1}$. Since the energy of states from each subspace $\tilde{\mathcal{H}}_p$ is measured on a different scale of $V(2(l-p)-1)$, the energy spectrum splits into bands corresponding to those subspaces. The energy gap between the pth and $(p+1)$st bands is of the order of $V(2(l-p)-1) - V(2(l-p-1)-1)$, and hence the largest gap is that between the 0th band and the 1st band, the next largest is that between the 1st band and 2nd band, and so on.

Figure 9.3 demonstrates on the example of four electrons to what extent this hidden symmetry holds for the Coulomb pseudopotential in the lowest Landau

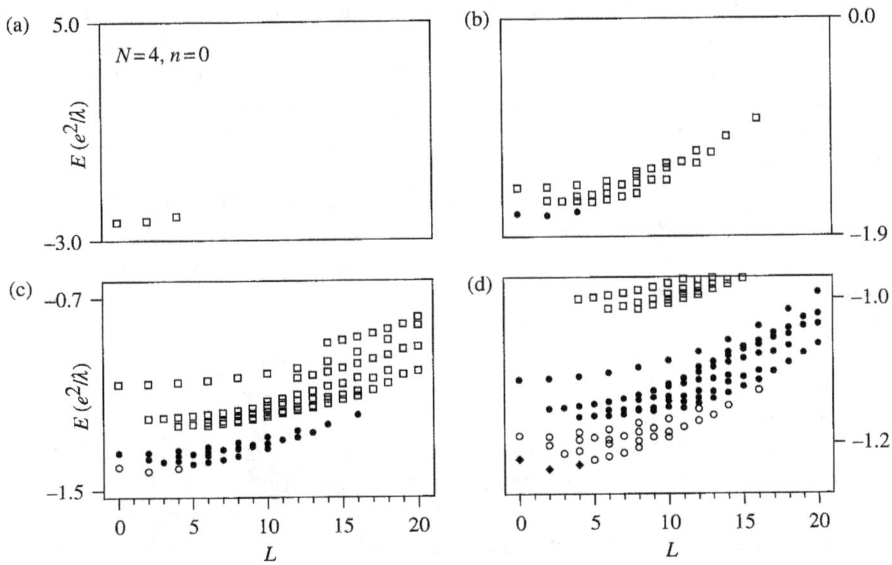

FIG. 9.3. Energy spectra of four electrons in the lowest Landau level each with angular momentum (a) $l = 5/2$, (b) $l = 11/2$, (c) $l = 17/2$, and (d) $l = 23/2$. Different subspaces $\mathcal{H}_p$ are marked with squares ($p = 0$), full circles ($p = 1$), open circles ($p = 2$), and diamonds ($p = 3$).

level. The subspaces $\mathcal{H}_p$ are identified by calculating composite fermion GPs of all states. They are not exact eigenspaces of the Coulomb interaction, but the mixing between different $\mathcal{H}_p$ is weak and the coefficients $\mathcal{G}(L')$ for $L' > 2(l-p)$ (which vanish exactly in exact subspaces $\mathcal{H}_p$) are indeed much smaller in states marked with a given p than in all other states. For example, for $2l = 11$, $\mathcal{G}(10) < 0.003$ for states marked with full circles, and $\mathcal{G}(10) > 0.1$ for all other states (squares).

Note that the set of angular momentum multiplets which form subspace $\tilde{\mathcal{H}}_p$ of N electrons each with angular momentum l is always the same as the set of multiplets in subspace $\tilde{\mathcal{H}}_{p+1}$ of N electrons each with angular momentum $l+(N-1)$. When l is increased by $N-1$, an additional band appears at high energy, but the structure of the low energy part of the spectrum is completely unchanged. For example, all three allowed multiplets for $l = 5/2$ ($L = 0$, 2, and 4) form the lowest energy band for $l = 11/2$, $17/2$, and $23/2$, where they span the $\tilde{\mathcal{H}}_1$, $\tilde{\mathcal{H}}_2$, and $\tilde{\mathcal{H}}_3$ subspace, respectively. Similarly, the first excited band for $l = 11/2$ is repeated for $l = 17/2$ and $23/2$, where it corresponds to $\tilde{\mathcal{H}}_1$ and $\tilde{\mathcal{H}}_2$ subspace, respectively.

Let us stress the fact that identical sets of multiplets occur in subspace $\tilde{\mathcal{H}}_p$ for a given l and in subspace $\tilde{\mathcal{H}}_{q+1}$ for l replaced by $l+(N-1)$, does not depend on the form of interaction, and follows solely from the rules of addition of angular momenta of identical fermions (this well-known empirical property has been recently rigorously proven using combinatorial methods (Benjamin *et al.* 2001)). However, if the interaction pseudopotential has short range, then: (a) $\tilde{\mathcal{H}}_p$ are interaction eigensubspaces; (b) energy bands corresponding to $\tilde{\mathcal{H}}_p$ with higher p lie below those of lower p; (c) spacing between neighbouring bands is governed by a difference between appropriate pseudopotential coefficients; and (d) wave functions and structure of energy levels within each band are insensitive to the details of interaction. Replacing V_{SR} by a pseudopotential that increases more slowly with increasing L' leads to: (e) coupling between subspaces $\tilde{\mathcal{H}}_p$; (f) mixing, overlap, or even order reversal of bands; (g) deviation of wave functions and the structure of energy levels within bands from those of the hard core repulsion (and thus their dependence on details of the interaction pseudopotential). The numerical calculations for the Coulomb pseudopotential in the lowest Landau level show (to a large extent) all short-range properties (a)–(d), and virtually no effects (e)–(g), characteristic of 'non SR' pseudopotentials.

The reoccurrence of L multiplets forming the low energy band when l is replaced by $l \pm (N-1)$ has the following crucial implication. In the lowest Landau level, the lowest energy (pth) band of the N electron spectrum at the monopole strength $2S$ contains L multiplets which are all the allowed N electron multiplets at $2S-2p(N-1)$. But $2S-2p(N-1)$ is just $2S^*$, the effective monopole strength of composite fermions! The mean field CS transformation which binds $2p$ fluxes (vortices) to each electron selects the same L multiplets from the entire spectrum as does the introduction of a hard core, which forbids a pair of electrons to be in a state with $L' > 2(l-p)$.

9.5 Definition of short-range pseudopotential

A useful operator identity relates total (L) and pair ($\hat{L}_{ij}$) angular momenta (Wójs and Quinn 1998, 1999, 2000c)

$$\sum_{i<j} \hat{L}_{ij}^2 = \hat{L}^2 + N(N-2)\hat{l}^2. \tag{9.10}$$

It implies that interaction given by a pseudopotential $V(L')$ that is linear in $\hat{L}'^2$ (e.g. the harmonic repulsion within each Landau level; see Eqn (9.4)) is degenerate within each L subspace and its energy is a linear function of $L(L+1)$. The many-body ground state has the lowest available L and is usually degenerate, while the state with maximum L has the largest energy. Note that this result is opposite to the Hund rule valid for spherical harmonics, due to the opposite behaviour of $V(L')$ for the fractional quantum Hall ($n=0$ and $l=S$) and atomic ($S=0$ and $l=n$) systems.

Deviations of $V(L')$ from a linear function of $L'(L'+1)$ lead to the level repulsion within each L subspace, and the ground state is no longer necessarily the state with minimum L. Rather, it is the state at a low L whose multiplicity N_L (number of different L multiplets) is large. It is interesting to observe that the L subspaces with relatively high N_L coincide with the mean field composite fermion prediction. In particular, for a given N, they reoccur at the same Ls when l is replaced by $l \pm (N-1)$, and the set of allowed Ls at a given l is always a subset of the set at $l+(N-1)$.

As we said earlier, if $V(L')$ has short range, the lowest energy states within each L subspace are those maximally avoiding large L', and the lowest band (separated from higher states by a gap) contains states in which a number of largest values of L' is avoided altogether. What is the criterion for the short range of $V(L')$? It follows from the following (Wójs 2001):

Theorem 9.1 *If for any three pair eigenstates at the pair angular momenta $L_1' < L_2' < L_3'$ the pseudopotential V decreases more quickly than linearly as a function of the average squared separation $\langle r^2 \rangle$, then the energy E_L of a many-electron state can be lowered without changing its total angular momentum L by transferring some of the parentage from $\mathcal{G}(L_1')$ and $\mathcal{G}(L_3')$ to $\mathcal{G}(L_2')$.*

This theorem was confirmed numerically (Wójs and Quinn 1998, 1999, 2000c), and it can be easily proven on a sphere by noticing that the above-mentioned transfer of (infinitesimal) parentage without changing L means replacing $\mathcal{G}(L_1')$, $\mathcal{G}(L_2')$, and $\mathcal{G}(L_3')$ by $\mathcal{G}(L_1')-\delta_1$, $\mathcal{G}(L_2')+\delta_2$, and $\mathcal{G}(L_3')-\delta_3$, respectively, such that $\delta_1+\delta_3=\delta_2$ and $\delta_1 L_1'(L_1'+1)+\delta_3 L_3'(L_3'+1)=\delta_2 L_2'(L_2'+1)$. Clearly, such a transfer does not change the total energy E_L given by Eqn (9.7) if V is harmonic (i.e. linear in $L'(L'+1)$), and that it decreases or increases E_L if V is super- or sub-harmonic, respectively.

It follows from Theorem 9.1 that if $V(\mathcal{R})$ is super-harmonic at large L' (small $\mathcal{R}$; i.e. at short range), the lowest energy states at each L will have minimum possible parentage from the (most strongly repulsive) pair state at the smallest value of $\mathcal{R} = 1$. Depending on the values of N and $2l$, the parentage from $\mathcal{R} = 1$ may even be avoided completely in the lowest energy states at some L. The complete avoidance of p pair states at $\mathcal{R} \leq m = 2p - 1$ is described by a Jastrow prefactor $\prod_{i<j}(z_i - z_j)^{2p}$ in the many-electron wave function. In particular, the Laughlin incompressible $\nu = 1/m$ ground state (Laughlin 1983*b*) is the only state at a given N and $2l$ for which $\mathcal{G}(L') = 0$ for all L' corresponding to $\mathcal{R} < m$.

Theorem 9.1 means that the so-called Laughlin correlations, in which the total parentage from the most strongly repulsive pair states is minimized in the lowest many-body states, occur for all pseudopotentials which (on a sphere) increase more quickly than linearly as a function of $L'(L'+1)$. For $V_\beta(L') = [L'(L'+1)]^\beta$, exponent $\beta > 1$ defines the class of short-range pseudopotentials, to which the mean field composite fermion picture can be applied. Within this class, the structure of low-lying energy spectrum and the corresponding wave functions very weakly depend on β and converge to those of V_{SR} for $\beta \to \infty$.

The extension of the short-range definition to $V(L')$ that are not strictly in the form of $V_\beta(L')$ is straightforward. If $V(L') > V(2l - m)$ for $L' > 2l - m$ and $V(L') < V(2l-m)$ for $L' < 2l-m$ and $V(L')$ increases more quickly than linearly as a function of $L'(L'+1)$ in the vicinity of $L' = 2l - m$, then pseudopotential $V(L')$ behaves like a short-range one at filling factors near $\nu = 1/m$.

9.6 Application to various pseudopotentials

It follows from Fig. 9.2(a) that the Coulomb pseudopotential in the lowest Landau level satisfies the short-range condition in the entire range of pair angular momentum L'; this is what validates the mean field composite fermion picture for filling factors $\nu \leq 1$. It also explains the formation of incompressible states of charged magnetoexcitons (X^-) formed in the electron–hole plasma (Wójs *et al.* 1998, 1999*a*). It has been shown that being charged fermions, the X^-s interact with one another through an effective pseudopotential that is similar to that of a pair of electrons and, in particular, has short range. Since the charged magnetoexcitons are formally equivalent to the finite-size skyrmions—the elementary charged excitations of the electron system at the filling factor $\nu = 1$ which carry massive spin—the latter are also expected to have Laughlin correlations. Even if the formation of an incompressible Laughlin liquid of any of these more complex objects (X^-s or skyrmions) were not possible (e.g. because of probable difficulties in creating their sufficiently high density, corresponding to $\nu = 1/3$, and at the same time keeping the system at low temperature), the general tendency of avoiding highly repulsive pair states means the absence of high-energy collisions between these charged particles, which may have significant consequences even at low density. For example, it is believed (Wójs and Quinn 2000*b*) that the absence of strong collisions between the so-called dark triplet X^-s and the surrounding charged particles (electrons or other X^-s) is responsible for the very weak radiative recombination of this complex, which is predicted to have infinite optical lifetime only in the translationally invariant system, that is, in the absence of the symmetry-breaking collisions.

A very different situation takes place in the excited Landau levels. The tendency for Laughlin correlations in a higher, nth Landau level is only true for $L' < 2(l-n)-1$ (see Fig. 9.2(b) for $n = 1$) and the mean field composite fermion picture is valid only for ν_n (filling factor in the nth Landau level) around and below $(2n+3)^{-1}$. Indeed, the mean field composite fermion features in the 10 electron energy spectra around $\nu = 1/3$ (in Fig. 9.1) are absent for the same fillings of the $n = 1$ Landau level (Wójs and Quinn 1998, 1999, 2000*c*).

One consequence of this is that the mean field composite fermion picture or Laughlin-like wave function cannot be used to describe the reported (Willett *et al.* 1987) incompressible state at $\nu = 2 + 1/3 = 7/3$ ($\nu_1 = 1/3$). The correlations in the $\nu = 7/3$ ground state are different than at $\nu = 1/3$; the origin of (apparent) incompressibility cannot be attributed to the formation of a Laughlin-like $\nu_1 = 1/3$ state (in which pair states with smallest average separation d^2 are avoided) on top of the $\nu = 2$ state, and connection between the excitation gap and the pseudopotential parameters is different. This is clearly visible in the dependence of the excitation gap Δ on the electron number N, plotted in Fig. 9.4 for $\nu = 1/3$ and $1/5$ fillings of the lowest and first excited Landau level. The gaps for $\nu = 1/5$ behave very similar to the function of N in both Landau levels, while it is not even possible to make a conclusive statement about degeneracy or incompressibility of the $\nu = 7/3$ state based on our data for up to 12 electrons.

More direct evidence that the Laughin correlations do not generally occur in the excited Landau level comes from comparison of the fractional parentage profiles, $\mathcal{G}(\mathcal{R})$, which can be regarded as a kind of pair-correlation function

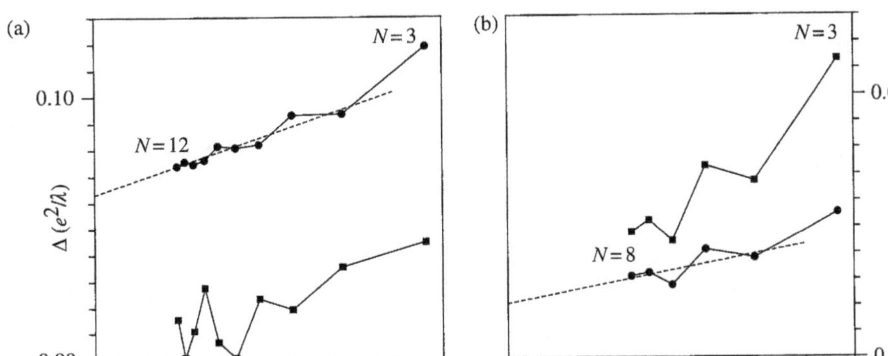

FIG. 9.4. Excitation gap Δ as a function of inverse electron number $1/N$ for filling factors (a) $\nu = 1/3$ and (b) $1/5$ in the $n = 0$ (circles) and $n = 1$ (squares) Landau levels.

(pair amplitude as a function of the relative orbital quantum number) for the low-lying many-body states at the same filling factor $\nu \sim 1/3$ of each Landau level. Such profiles for the lowest $L = 0$ state in the Coulomb spectrum of $N = 12$ or 14 electrons at three different values of $2l$ are compared for the lowest, first excited, and second excited Landau level in Fig. 9.5. Clearly, the Laughlin 'correlation hole' at $\mathcal{R} = 1$ characteristic of Laughlin-correlated low-lying states in the lowest Landau level does not occur in the excited Landau levels.

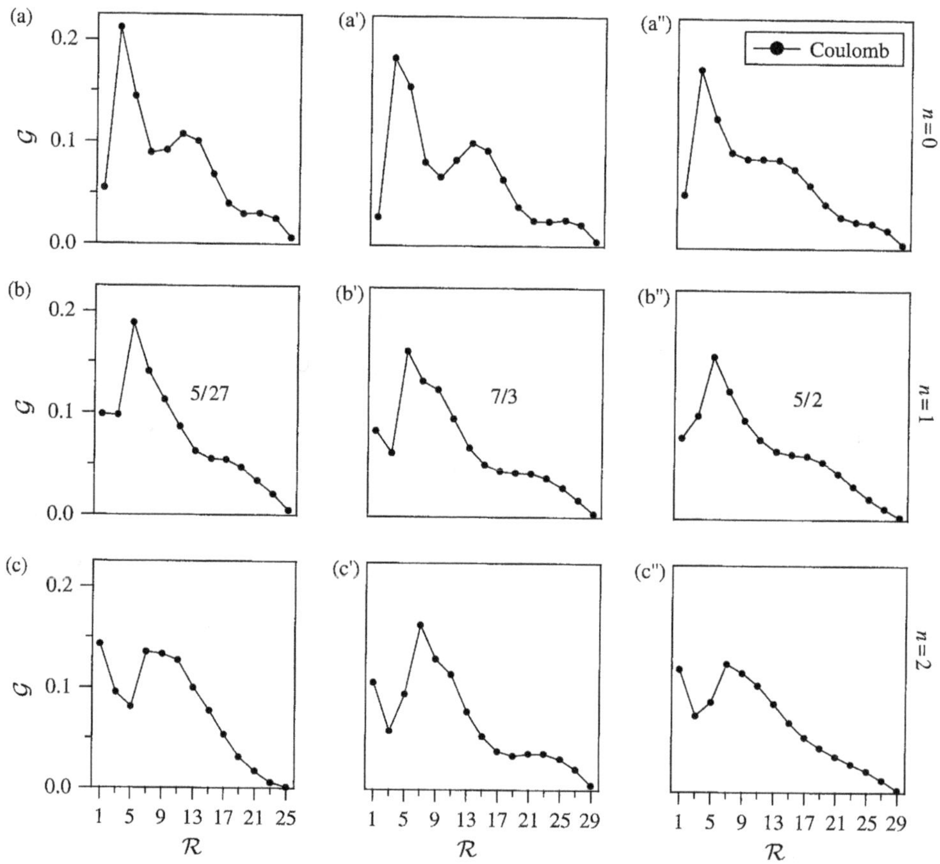

FIG. 9.5. The pair-correlation functions (coefficient of fractional parentage $\mathcal{G}$ versus relative pair angular momentum $\mathcal{R}$) in the lowest energy $L = 0$ state of N electrons on a Haldane's sphere: (a–c) $N = 12$ and $2l = 25$, (a′–c′) $N = 12$ and $2l = 29$, and (a″–c″) $N = 14$ and $2l = 29$, calculated for the Coulomb pseudopotential in the (a–a″) lowest, (b–b″) first excited, and (c–c″) second excited Landau level.

The short-range criterion can be applied to the QP pseudopotentials and it can be seen that QPs do not form incompressible states at all Laughlin filling factors $\nu_{\rm QP} = 1/m$ in the hierarchy picture (Haldane 1983; Sitko *et al.* 1997*b*) of fractional quantum Hall states. Lines in Fig. 9.1(b) and (f) mark $V_{\rm QE}$ and $V_{\rm QH}$ for the Laughlin $\nu = 1/3$ state of 10 electrons. Clearly, the incompressible states with a large gap will be formed by QHs at $\nu_{\rm QH} = 1/3$ and by QEs at $\nu_{\rm QE} = 1$, explaining strong FQHE of the underlying electron system at Jain $\nu = 2/7$ and $2/5$ fractions, respectively. On the other hand, there is no indication for condensed states at $\nu_{\rm QH} = 1/5$ ($\nu = 4/13$) or $\nu_{\rm QE} = 1/3$ ($\nu = 4/11$), and the gap above possibly incompressible $\nu_{\rm QH} = 1/7$ ($\nu = 6/19$) and $\nu_{\rm QE} = 1/5$ ($\nu = 6/17$) is very small. However, recent experimental findings indicate the possibility for odd-denominator 4/11, 4/13, 6/17, and also 5/13, 5/17, 7/11 hierarchical states—see Fig. 9.6.

The pseudopotentials for the interaction between the QEs and between the QHs in the $\nu = 1/3$ and $1/5$ Laughlin (parent) states are plotted in Fig. 9.7. The data shown comes from the numerical spectra similar to that of Fig. 9.1 but obtained for different electron numbers N. Clearly, all curves converge to the limiting behaviour characteristic for an infinite (planar) system.

Similar analysis has been carried out for the interaction between the reversed-spin quasiparticle excitations of the Laughlin liquid, so-called reversed-spin quasielectrons ($\rm QE_R$) first discovered by Rezayi (1987). The energy spectrum corresponding to two QEs in the Laughlin $\nu = 1/3$ ground state, similar to those in Fig. 9.1(b) but also including states with only partial spin (J) polarization, is presented in Fig. 9.8(a). In addition to the QE–QE band identified earlier, the many-body states containing QE–$\rm QE_R$ and $\rm QE_R$–$\rm QE_R$ pairs occur, from which the pseudopotentials describing the QE–$\rm QE_R$ and $\rm QE_R$–$\rm QE_R$ repulsion can be

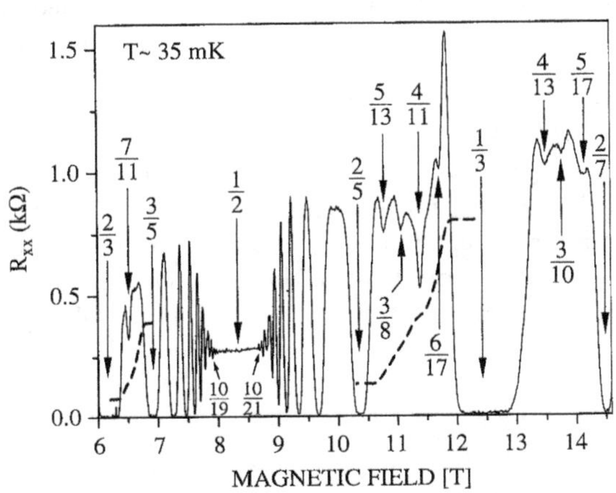

FIG. 9.6. Indications for the quantum Hall states at odd-denominator hierarchical fractions 4/11, 4/13, 5/13, 6/17, 5/17, 7/11 (from Pan *et al.* (2003)), and even-denominator hierarchical fractions 3/8 and 3/10—see Section 8.3.

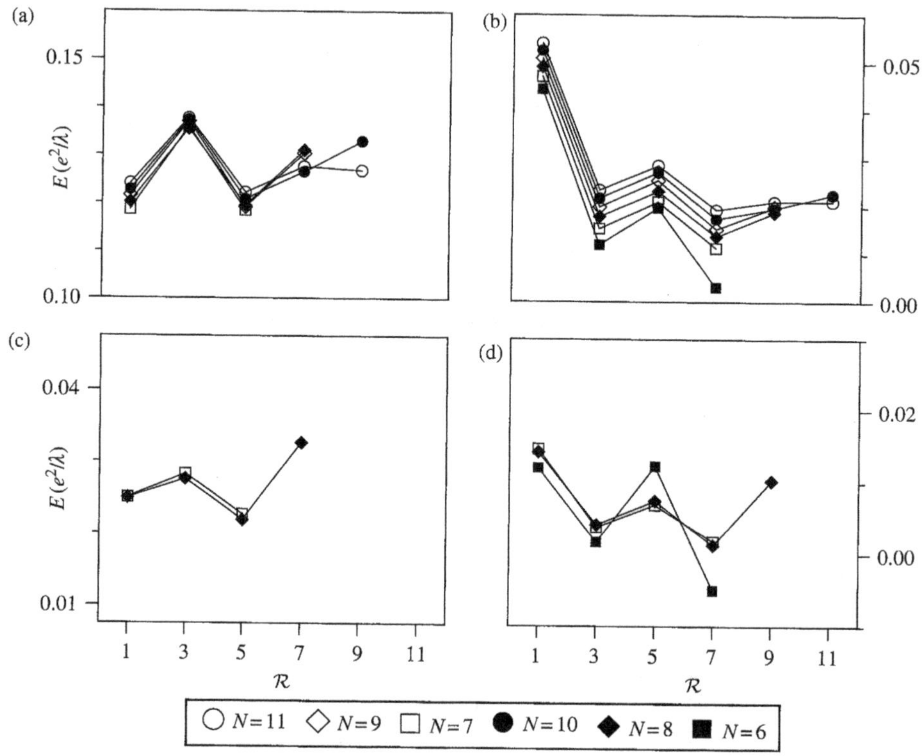

FIG. 9.7. Energies $E = 2\varepsilon + V(\mathcal{R})$ of a pair of (a and c) quasielectrons and (b and d) quasiholes in Laughlin $\nu = 1/3$ and $\nu = 1/5$ states, respectively, as a function of relative pair angular momentum $\mathcal{R}$, obtained in diagonalization of N electrons.

calculated. The $\mathrm{QE_R}$–$\mathrm{QE_R}$ pseudopotentials calculated for $N = 6$–9 electrons are drawn in Fig. 9.8(b). Clearly, it seems that they all have super-harmonic character, suggesting incompressibility of the corresponding partially spin-unpolarized hierarchy states. The most prominent of these states, the $\nu = 1/3$ state of $\mathrm{QE_R}$s in the $\nu = 1/3$ (parent) state of electrons, occurs at $\nu = 4/11$.

9.7 Multi-component systems

For the (one-component) electron gas on a plane, avoiding pair states with $\mathcal{R} < m$ is achieved with the factor $\prod_{i<j}(x_i - x_j)^m$ in the Laughlin $\nu = 1/m$ wave function. For a system containing a number of distinguishable types of fermions interacting through Coulomb-like pseudopotentials, the appropriate generalization of the Laughlin wave function will contain a factor $\prod(x_i^{(a)} - x_j^{(b)})^{m_{ab}}$, where

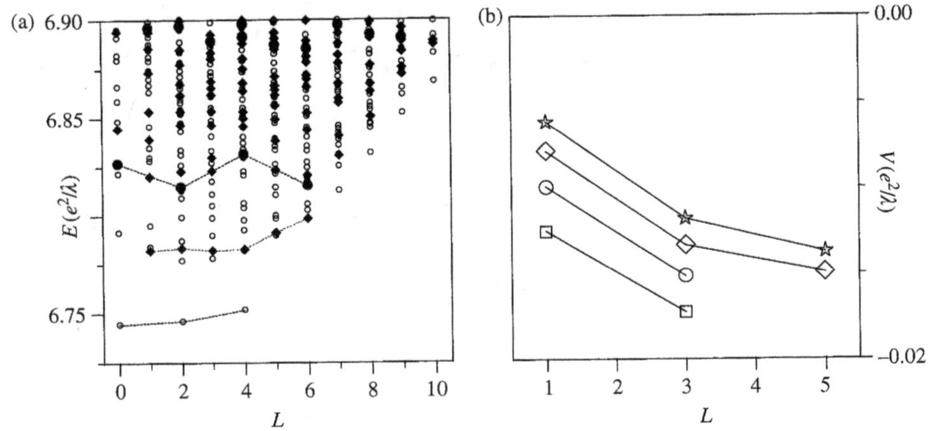

FIG. 9.8. (a) The energy spectrum (Coulomb energy E versus angular momentum L) of the system of $N = 8$ electrons on Haldane sphere at the monopole strength $2S = 3(N-1) - 2 = 19$. Filled circles, diamonds, and open circles mark states with the total spin $J = 1/2N = 4$ (maximum polarization), $J = 1/2N - 1 = 3$ (one reversed spin), and $J = 1/2N - 2 = 2$ (two reversed spins), respectively. Lines connect states containing one QE–QE ($J = 4$), QE–QE$_\mathrm{R}$($J = 3$), or QE$_\mathrm{R}$–QE$_\mathrm{R}$ ($J = 2$) pair. (b) The pseudopotentials (pair energy V versus relative angular momentum $\mathcal{R}$) of the QE$_\mathrm{R}$–QE$_\mathrm{R}$ interaction calculated in the systems of $N \le 9$ electrons on Haldane sphere. λ is the magnetic length.

$x_i^{(a)}$ is the complex coordinate for the position of ith particle of type a, and the product is taken over all pairs. For each type of particle one power of $(x_i^{(a)} - x_j^{(a)})$ results from the antisymmetrization required for indistinguishable fermions and the other factors describe Jastrow-type correlations between the interacting particles. Such a wave function guarantees that $\mathcal{R}_{ab} \ge m_{ab}$, for all pairings of various types of particles, thereby avoiding large pair repulsion (Halperin 1983; Haldane and Rezayi 1988). Fermi statistics of particles of each type requires that all m_{aa} are odd, and the hard cores (due to Pauli's exclusion principle when applied to constituent fermions: electrons and holes) require that $m_{ab} \ge \mathcal{R}_{ab}^{\min}$ for all pairs.

An example of the physical system in which two types of charged fermions occur is an electron–hole plasma in the lowest Landau level (obtained by optical excitation of the electron system), where part of the electrons bind to the small number of holes to form the charged magnetoexcitons, X^-, and the remaining electrons stay unbound. Each X^- consists of two electrons and one hole, and thus it is a charged fermion. Larger excitonic ions (such as $X_2^- = 3e + 2h$, $X_3^- = 4e + 3h$, etc.) are also possible in an electron–hole plasma but their binding energy is smaller than that of the X^-.

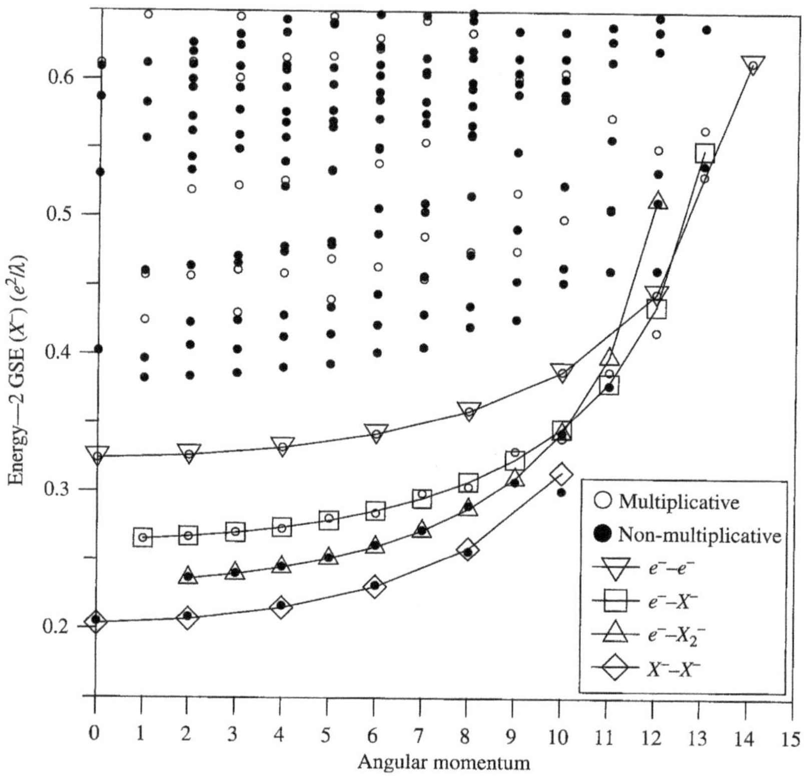

FIG. 9.9. Energy spectrum of four electrons and two holes at $2S = 15$. Open circles—multiplicative states; solid circles—non-multiplicative states; triangles, squares, and diamonds—approximate pseudopotentials.

The pseudopotentials describing interaction between electrons and different excitonic ions can be obtained from the energy spectra of small systems containing a few holes and two more electrons. In Fig. 9.9 we display the energy spectrum for four electrons and two holes at $2S = 15$. The states marked by open and solid circles are the multiplicative states (containing one or more decoupled excitons, Xs) and the non-multiplicative states, respectively. The e^-–e^-, e^-–X^-, e^-–X_2^-, and $X^- - X^-$ pseudopotentials identified as indicated are all repulsive and all have short range.

The typical energy spectra of a larger electron–hole ($8e + 2h$) system on a Haldane sphere is shown in Fig. 9.10. The monopole strength is $2S = 9$, 13, and 14 in different frames. Filled circles mark the non-multiplicative states, and the open circles and squares mark the multiplicative states (states with one and two decoupled excitons), respectively. In frames (b), (d), and (f) we plot the low energy spectra of different charge complexes interacting through appropriate effective pseudopotentials, corresponding to four possible groupings of bound excitonic ions and unbound electrons: (i) $4e^- + 2X^-$, (ii) $5e^- + X_2^-$, (iii) $5e^- + X^- + X^0$, and (iv) $6e^- + 2X^0$. By comparing left and right frames, we can identify low-lying states containing all different groupings in the electron–hole spectra.

In order to understand the numerical results presented in Fig. 9.10, it is useful to introduce a generalized composite fermion picture (Wójs *et al.* 1999*b*) by attaching to each particle fictitious flux tubes carrying an integral number of flux quanta ϕ_0. In the multi-component system, each a-particle carries flux $(m_{aa} - 1)\phi_0$ that couples only to charges on all other a-particles and fluxes $m_{ab}\phi_0$ that couple only to charges on all b-particles, where a and b are any of the types of fermions. The effective monopole strength (Jain 1989; Chen and Quinn 1994; Sitko *et al.* 1996; Wójs and Quinn 1998, 1999, 2000*c*) seen by a composite fermion of type a (composite fermion-a) is

$$2S_a^* = 2S - \sum_b (m_{ab} - \delta_{ab})(N_b - \delta_{ab}). \tag{9.11}$$

For different multi-component systems we expect generalized Laughlin incompressible states (for two components denoted as (m_{AA}, m_{BB}, m_{AB})) when all the hard core pseudopotentials are avoided and composite fermions of each kind fill completely an integral number of their composite fermion shells (e.g. $N_a = 2l_a^* + 1$ for the lowest shell). In other cases, the low lying multiplets are expected to contain different kinds of quasiparticles (QP-A, QP-B, ...) or quasiholes (QH-A, QH-B, ...) in the neighbouring incompressible state.

Our multi-component composite fermion picture can be applied to the system of excitonic ions, where the composite fermion angular momenta are given by $l^*_{X_k^-} = |S^*_{X_k^-}| - k$. As an example, let us analyse the low lying $8e + 2h$ states in Fig. 9.10. At $2S = 9$, for $m_{e^-e^-} = m_{X^-X^-} = 3$ and $m_{e^-X^-} = 1$, we predict the following low-lying multiplets in each grouping: (i) $2S^*_{e^-} = 1$ and $2S^*_{X^-} = 3$ gives $l^*_{e^-} = l^*_{X^-} = 1/2$. Two composite fermion-$X^-$s fill their lowest shell ($L_{X^-} = 0$) and we have two QP-e^-s in their first excited shell, each with angular momentum $l^*_{e^-} + 1 = 3/2$ ($L_{e^-} = 0$ and 2). Addition of L_{e^-} and L_{X^-} gives total angular momenta $L = 0$ and 2. We interpret these states as those of two QP-es in the incompressible [331] state. Similarly, for other groupings we obtain: (ii) $L = 2$; (iii) $L = 1$, 2, and 3; and (iv) $L = 0$ ($\nu = 2/3$ state of six electrons).

At $2S = 13$ and 14 we set $m_{e^-e^-} = m_{X^-X^-} = 3$ and $m_{e^-X^-} = 2$ and obtain the following predictions. First, at $2S = 13$: (i) the ground state is the incompressible [332] state at $L = 0$; the first excited band should therefore contain states with one QP–QH pair of either kind. For the e^- excitations, the QP-e^- and QH-e^- angular momenta are $l^*_{e^-} = 3/2$ and $l^*_{e^-} + 1 = 5/2$, respectively, and the allowed pair states have $L_{e^-} = 1$, 2, 3, and 4. However, the $L = 1$ state has to be discarded, as it is known to have high energy in the one-component (four electron) spectrum (Sitko *et al.* 1996). For the X^- excitations, we have $l^*_{X^-} = 1/2$ and pair states can have $L_{X^-} = 1$ or 2. The first excited band is therefore expected to contain multiplets at $L = 1$, 2^2, 3, and 4. The low lying

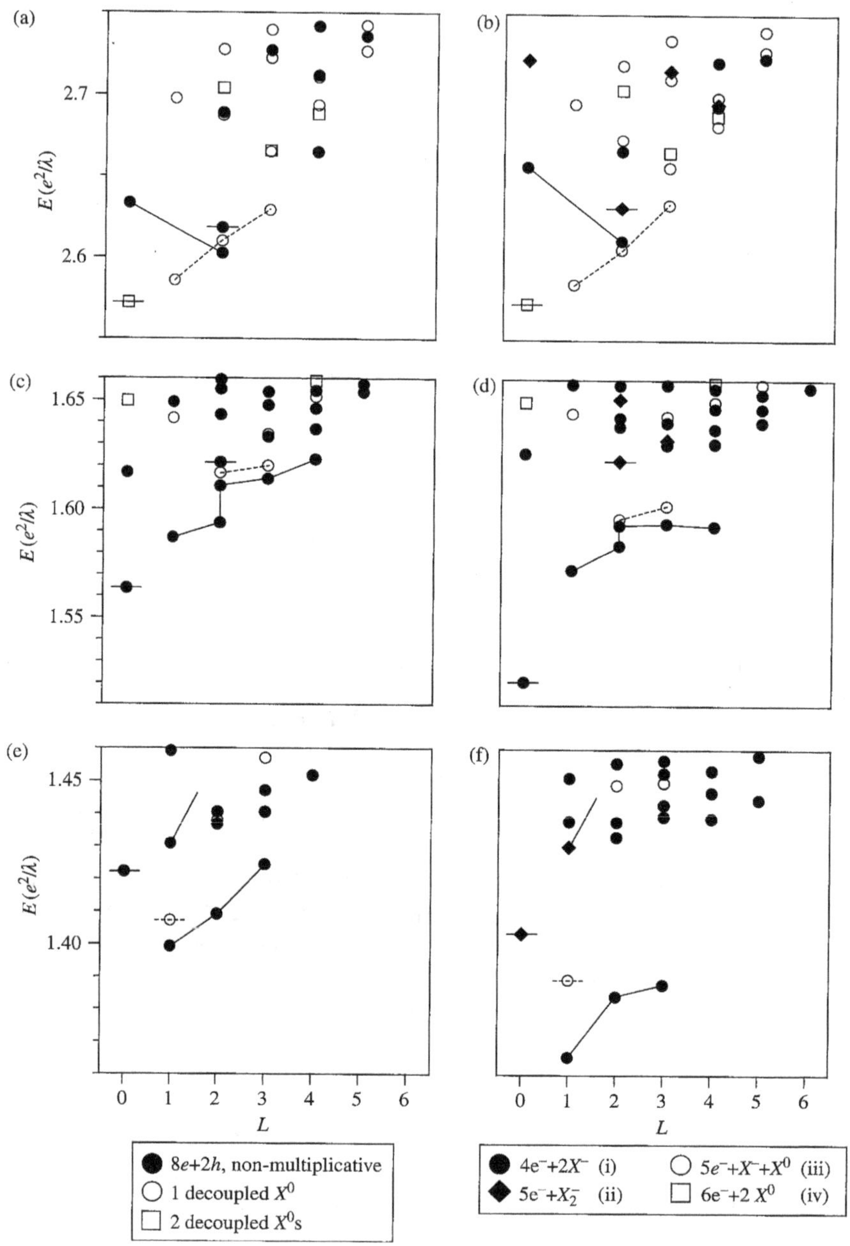

FIG. 9.10. Left: low energy spectra of the $8e + 2h$ system on a Haldane sphere at (a) $2S = 9$, (c) $2S = 13$, and (e) $2S = 14$. Right: approximate spectra calculated for all possible groupings containing excitons (charged composite particles interacting through effective pseudopotentials). Lines connect corresponding states in left and right frames.

multiplets for other groupings are expected at: (ii) $L = 2$ and 3; (iii) $2S^*_{X_2^-} = 3$ gives no bound X_2^- state; setting $m_{e^-X^-} = 1$ we obtain $L = 2$; and (iv) $L = 0$, 2, and 4. Finally, at $2S = 14$, we obtain: (i) $L = 1$, 2, and 3; (ii) incompressible [3*2] state at $L = 0$ ($m_{X^-X^-}$ is irrelevant for one X^-) and the first excited band at $L = 1$, 2, 3, 4, and 5; (iii) $L = 1$; and (iv) $L = 3$.

The agreement of our composite fermion predictions with the data in Fig. 9.10 is really quite remarkable and strongly indicates that our multi-component composite fermion picture is correct. We were indeed able to confirm predicted Jastrow-type correlations in the low-lying states by calculating their coefficients of fractional parentage (de Shalit and Talmi 1963; Cowan 1981; Wójs and Quinn 1998, 1999, 2000*c*). We have also verified the composite fermion predictions for other systems that we were able to treat numerically. If exponents m_{ab} are chosen correctly, the composite fermion picture works well in all cases.

The authors acknowledge support by KBN grant No. 2 P03B 024 24.

APPENDIX A

HOMOTOPY GROUPS

Let Ω be a path-connected space in which any two points can be connected by a continuous curve in this space. We define the loops at the point ω_0 in the space Ω as closed continuous curves (with a direction of winding being defined) which come through the point ω_0. They can be treated as continuous transformations $\mathcal{F}(t)$ of the interval $[0,1]$ into the space Ω, in accord with the condition $\mathcal{F}(0) = \mathcal{F}(1) = \omega_0$. If $\mathcal{F}(t) = \omega_0$ for each $t \in [0,1]$, then one has a loop which is called a zero-loop.

The two loops $\mathcal{F}$ and $\mathcal{G}$ are homotopic when there is a family of transformations $\mathcal{H}_l(t)$ of the interval $[0,1]$ into the space Ω ($\mathcal{H}_l(t)$ is a continuous function for both l and t for $l \in [0,1]$), and the following conditions are fulfilled:

$$\begin{aligned} &(i) \quad \mathcal{H}_0(t) = \mathcal{F}(t), \\ &(ii) \quad \mathcal{H}_1(t) = \mathcal{G}(t), \\ &(iii) \quad \mathcal{H}_l(0) = \mathcal{H}_l(1) = \omega_0, \end{aligned}$$

for each $t \in [0,1]$ and $l \in [0,1]$. The examples of homotopic and non-homotopic loops are shown in Fig. A.1. A path-connected space in which each loop is homotopic with a zero-loop is called a simple-connected space.

A product of two loops, $\mathcal{F}$ and $\mathcal{G}$, is found when the loop $\mathcal{F}$ is described first, and after that the loop $\mathcal{G}$:

$$\mathcal{F} \cdot \mathcal{G} = \begin{cases} \mathcal{F}(2t), & 0 \leq t \leq 0.5, \\ \mathcal{G}(2t-1), & 0.5 < t \leq 1. \end{cases}$$

When describing a loop, the direction of winding is important. A class of loops $[\mathcal{F}]$ is understood as a set of all loops homotopic with the loop $\mathcal{F}$. Consequently, the set of all loops at ω_0 divides into classes of mutually homotopic loops (an individual loop $\mathcal{F}$ is called a representation of the class $[\mathcal{F}]$). A product of two classes of homotopic loops can be defined similarly as the product of loops:

$$[\mathcal{F}] \cdot [\mathcal{G}] \stackrel{def}{=} [\mathcal{F} \cdot \mathcal{G}].$$

It is important to say that the class of products of loops $[\mathcal{F} \cdot \mathcal{G}]$ does not depend on the choice of the representation of classes $[\mathcal{F}]$ and $[\mathcal{G}]$. One can check that classes with such a product make up a group (Mermin 1979; Duda 1991). Such a group is called a fundamental group of the space Ω at the point ω_0 or the first homotopy group, and the group is denoted by $\pi_1(\Omega, \omega_0)$. When a path-connected space is simple-connected, then its fundamental group becomes trivial, that is, it consists

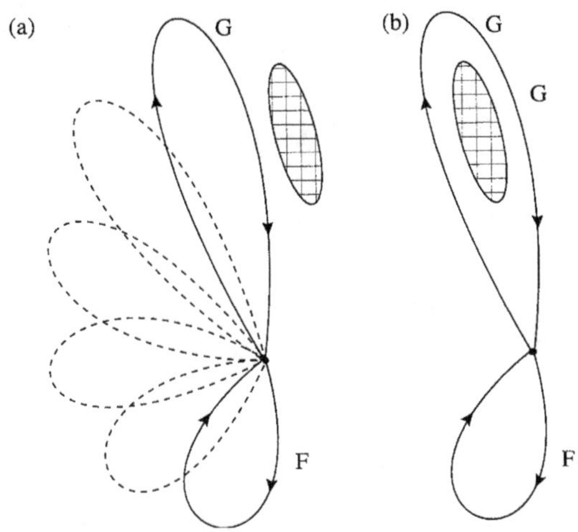

FIG. A.1. Examples of loops in the plane with a cut-out hole: (a) homotopic and (b) non-homotopic. In (b) loops $\mathcal{F}$ and $\mathcal{G}$ are non-homotopic; the loop $\mathcal{F}$ is a zero-loop and a hole is enclosed within the loop $\mathcal{G}$.

of only a single element. The distinguished point ω_0 in the space Ω is called a base point. The class of the zero-loop constitutes the identity element of the group $\pi_1(\Omega, \omega_0)$, and an inverse element of a class of a loop $[\mathcal{F}]$ is $[\mathcal{F}]^{-1} = [\mathcal{F}^{-1}]$, where the loop $\mathcal{F}^{-1}$ is oppositely wound with respect to the loop $\mathcal{F}$.

If one takes a continuous curve $\mathcal{C}$ connecting points ω_0 and ω_1 in the path-connected space Ω, then the transformation:

$$\pi_1(\Omega, \omega_0) \to \pi_1(\Omega, \omega_1) : [\mathcal{F}] \to [\mathcal{C}] \cdot [\mathcal{F}] \cdot [\mathcal{C}]^{-1}$$

is an isomorphism. Figure A.2 shows how to construct this isomorphism. From the above remarks one can note that the fundamental group of a path-connected space does not depend on the choice of a base point. Therefore, for path-connected spaces, the base point does not have to be defined and there exists a fundamental group or a homotopy group (this group is denoted as $\pi_1(\Omega)$).

We present now the theorems which are helpful in determining the fundamental groups for a more complex space. The fundamental group for the Carthesian product of path-connected spaces equals the simple sum of fundamental groups of these spaces

$$\pi_1(\Omega_1 \times \Omega_2) = \pi_1(\Omega_1) \oplus \pi_1(\Omega_2).$$

Since $\pi_1(R) = \mathcal{E}$, one has $\pi_1(R^n) = \mathcal{E}$ ($\mathcal{E}$—one-element group). Similarly, the fundamental group for the circle $\pi_1(S^1) = \mathcal{Z}$ allows to determine the first homotopy group for the torus $T = S^1 \times S^1$ in the form $\pi_1(T) = \pi_1(S^1) \oplus \pi_1(S^1) = \mathcal{Z} \oplus \mathcal{Z}$.

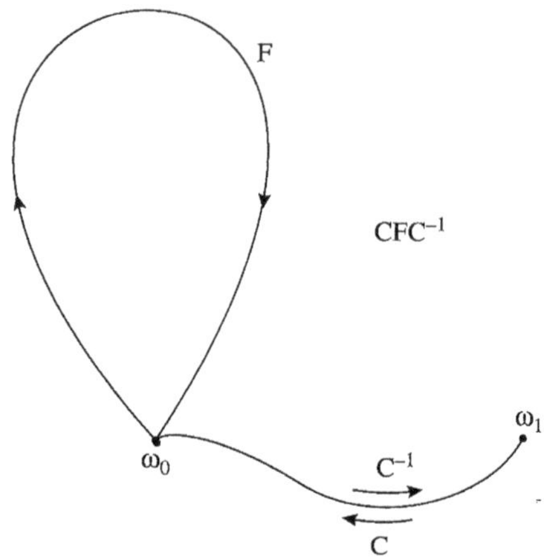

FIG. A.2. A construction of a loop $\mathcal{C} \cdot \mathcal{F} \cdot \mathcal{C}^{-1}$ at ω_1.

The following definition of the fundamental group can be extended on higher homotopy groups. The loop $\mathcal{F}$ can be defined as a transformation of the interval $[0,1]$ into the space Ω with connected ends $f(0) = f(1) = \omega_0$. Consequently, a loop can be obtained as a transformation $f : S^1, s \to \Omega, \omega_0$ of a circle into the space Ω and the point $\omega_0 \in \Omega$ constitutes an image of a selected point $s \in S^1$. This leads to the definition of the fundamental group as the set of classes of homotopies of all transformations $f : S^1, s \to \Omega, \omega_0$. Now a concept of the nth homotopy group can be introduced as a natural generalization of the fundamental group: the nth homotopy group (or the nth dimensional homotopy group) is the set of classes of homotopies of all transformations $f : S^n, s \to \Omega, \omega_0$ for $n = 0, 1, 2, \ldots$ This group is defined as $\pi_n(\Omega, \omega_0)$. The above definition is clear for $n > 0$; for $n = 0$ a set $\pi_0(\Omega, \omega_0)$, not a group, is obtained. The group $\pi_0(\Omega, \omega_0)$ is the set of path-components[1] of the space Ω with the point ω_0 as the base point. For a path-connected space, the zero-homotopy group becomes the one-element group. In certain circumstances in the zero-homotopy group, the group operation can be defined and the group structure can be introduced (Mermin 1979).

[1]Path-components of the space Ω are maximal path-connected subspaces of this space.

APPENDIX B

CORRELATION FUNCTION FOR AN ANYON GAS IN THE SELF-CONSISTENT HARTREE APPROXIMATION

The irreducible self-energy in the Hartree approximation constitutes a local single-particle potential which defines eigenstates fulfilling the self-consistent Schrödinger equation:

$$-\frac{\hbar^2\nabla_1^2}{2m}\phi_j(\mathbf{r}_1) + \int d\mathbf{r}_2 \hbar\Sigma_H(\mathbf{r}_1, \mathbf{r}_2)\phi_j(\mathbf{r}_2) = \hbar\omega_j\phi_j(\mathbf{r}_1). \tag{B.1}$$

The one-particle Green function can be expressed as:

$$G(1, 1') = \int \frac{d\omega}{2\pi} G(\mathbf{r}_1, \mathbf{r}_1', \omega) e^{-i\omega(t_1 - t_1')} \tag{B.2}$$

and making use of the Lehman representation:

$$G(\mathbf{r}_1, \mathbf{r}_1', \omega) = \sum_j \phi_j(\mathbf{r}_1)\phi_j^*(\mathbf{r}_1') \left[\frac{\theta(\omega_j - \omega_{\mathrm{F}})}{\omega - \omega_j + i\eta} + \frac{\theta(\omega_{\mathrm{F}} - \omega_j)}{\omega - \omega_j - i\eta}\right] \tag{B.3}$$

($\hbar\omega_{\mathrm{F}}$—Fermi energy). Let us introduce the projector on the nth Landau level, Π_n, then

$$G(\mathbf{r}_1, \mathbf{r}_1', \omega) = \sum_n \frac{\Pi_n(\mathbf{r}_1, \mathbf{r}_1')}{\omega - \omega_n + i\eta_n}. \tag{B.4}$$

Positive values of η_n correspond to filled up Landau levels, negative values of η_n to empty levels. In the Eqn (5.84), the value

$$-iG(8, 8'^{+}) = \sum_{n=0}^{f-1} \Pi_n(\mathbf{r}_8, \mathbf{r}_8') \tag{B.5}$$

appears, in which summation goes over filled Landau levels and for equal space variables $iG(8, 8^+) = \bar{\rho}(\mathbf{r}_8)$. Substituting the irreducible self-energy in the Hartree approximation (5.84) into the Schrödinger equation (B.1), the following expression is obtained

$$\begin{aligned} &\frac{1}{2m}\left(\mathbf{p}_1 + \frac{e}{c}\bar{\mathbf{A}}_1\right)^2 \phi_j(\mathbf{r}_1) \\ &\quad + \frac{e}{mc}\int d\mathbf{r}_3 \left[\mathbf{A}_{13} \cdot \left(\mathbf{p}_3 + \frac{e}{c}\bar{\mathbf{A}}_3\right)\langle\Psi^+(\mathbf{r}_3')\Psi(\mathbf{r}_3)\rangle\right]\Big|_{\mathbf{r}_3'=\mathbf{r}_3} \phi_j(\mathbf{r}_1) \\ &= \hbar\omega_j\phi_j(\mathbf{r}_1). \end{aligned} \tag{B.6}$$

The first term in the above expression denotes exactly mean field Hamiltonian H_0 and the second term disappears. Consequently, eigenstates of the system in the Hartree approximation are exactly Landau states and the self-consistent Green function in the Hartree approximation becomes the Green function of the unperturbed state, G^0. Let the vertex function in the Hartree approximation (5.92) be split up in the following way:

$$W_{\rm H}(\mathbf{r}_3\mathbf{r}_4,\mathbf{r}_5\mathbf{r}_6) = W_{\rm H}^1(\mathbf{r}_3\mathbf{r}_4,\mathbf{r}_5\mathbf{r}_6) + W_{\rm H}^2(\mathbf{r}_3\mathbf{r}_4,\mathbf{r}_5\mathbf{r}_6), \tag{B.7}$$

where

$$W_{\rm H}^1(\mathbf{r}_3\mathbf{r}_4,\mathbf{r}_5\mathbf{r}_6) = w_0(\mathbf{r}_{35})\delta(\mathbf{r}_3-\mathbf{r}_4)\delta(\mathbf{r}_5-\mathbf{r}_6), \tag{B.8}$$

$$\begin{aligned} W_{\rm H}^2(\mathbf{r}_3\mathbf{r}_4,\mathbf{r}_5\mathbf{r}_6) = \frac{e}{mc}\iint d\mathbf{r}_7\,d\mathbf{r}_8 &\left[\mathbf{A}_{78}\cdot\left(\mathbf{p}_7+\frac{e}{c}\bar{\mathbf{A}}_7-\mathbf{p}_8-\frac{e}{c}\bar{\mathbf{A}}_8\right)\right.\\ &\left.\times\;\delta(\mathbf{r}_5-\mathbf{r}_8)\delta(\mathbf{r}_6-\mathbf{r}_8')\delta(\mathbf{r}_4-\mathbf{r}_7)\delta(\mathbf{r}_3-\mathbf{r}_7')\right]\Big|_{\mathbf{r}_7'=\mathbf{r}_7,\mathbf{r}_8'=\mathbf{r}_8}. \end{aligned} \tag{B.9}$$

The two-particle correlation function depends only on the difference of coordinates t_1-t_2, thus

$$F(11',2'2) = \int\frac{d\omega}{2\pi}F(\mathbf{r}_1\mathbf{r}_1',\mathbf{r}_2\mathbf{r}_2',\omega)e^{-i\omega(t_1-t_2)}. \tag{B.10}$$

The same relationship refers to unperturbed function F_0 which simplifies the Bethe–Salpeter equation (5.90). As the self-consistent one-particle Green function in the Hartree approximation becomes the Green function of unperturbed state, then

$$F_0(\mathbf{r}_1\mathbf{r}_1',\mathbf{r}_2'\mathbf{r}_2,\omega) = -i\int\frac{d\omega'}{2\pi}G^0(\mathbf{r}_1,\mathbf{r}_2',\omega+\omega')G^0(\mathbf{r}_2,\mathbf{r}_1',\omega'). \tag{B.11}$$

The above function constitutes the time-ordered correlation function of the unperturbed system. However, to study the linear response of the system, an information about the retarded correlation function is needed, given by the following expression:

$$\begin{aligned} F_0(\mathbf{r}_1\mathbf{r}_1',\mathbf{r}_2'\mathbf{r}_2,\omega) = &\sum_{m,n}\Pi_m(\mathbf{r}_1,\mathbf{r}_2')\Pi(\mathbf{r}_2,\mathbf{r}_1')\\ &\times\left[\frac{\theta(\omega_m^0-\omega_{\rm F}^0)\theta(\omega_{\rm F}^0-\omega_n^0)}{\omega-(\omega_m^0-\omega_n^0)+i\eta}-\frac{\theta(\omega_n^0-\omega_{\rm F}^0)\theta(\omega_F^0-\omega_m^0)}{\omega+(\omega_n^0-\omega_m^0)+i\eta}\right]. \end{aligned} \tag{B.12}$$

The magnetoexciton basis serves well for detailed calculations. For two particles let us define coordinates of the centre of mass $\mathbf{R}=\frac{1}{2}(\mathbf{r}_1+\mathbf{r}_2)$, and the relative coordinates $\mathbf{r} = \mathbf{r}_1 - \mathbf{r}_2$. Then the Hamiltonian of a hole and a particle assumes the form:

$$H = \frac{1}{2m}\left[\frac{1}{2}\left(\mathbf{P} + \frac{e}{2c}\mathbf{B}^s \times \mathbf{r}\right)^2 + 2\left(\mathbf{p} + \frac{e}{2c}\mathbf{B}^s \times \mathbf{R}\right)^2\right], \tag{B.13}$$

where $\mathbf{P} = \mathbf{p}_1 + \mathbf{p}_2$, $\mathbf{p} = \frac{1}{2}(\mathbf{p}_1 - \mathbf{p}_2)$. One can note that for such a system of two particles, the following 'momentum' operator

$$\mathbf{Q} = \mathbf{P} - \frac{e}{2c}\mathbf{B} \times \mathbf{r} \tag{B.14}$$

commutes with the Hamiltonian. Consequently, an operator $\mathbf{Q}(\mathbf{q})$ eigenvalue is a good quantum number. In the magnetoexciton basis, states are numbered by two numbers $M = n - m$, $N = \min(n, m)$, while n, m number single-particle Landau states of both particles. Introducing complex coordinates of a position $z_j = x_j + iy_j$ ($j = 1, 2$) and a 'momentum' $z_\alpha = iq_x - q_y$, the two-particle magnetoexciton wave function can be expressed as ($a_0 = 1$) (Fetter and Hanna 1992*a*):

$$\begin{aligned}\Psi_{NM\alpha}(\mathbf{r}_1, \mathbf{r}_2) = &\left[\frac{N!}{2\pi L^2 2^{|M|}(N + |M|)!}\right]^{1/2} \gamma_M(z_1 - z_2 - z_\alpha)\\ &\times L_N^{|M|}\left(\frac{1}{2}|z_1 - z_2 - z_\alpha|^2\right) \exp\left[-\frac{1}{4}(|z_1|^2 + |z_2|^2 + |z_\alpha|^2)\right]\\ &\times \exp\left[\frac{1}{2}(z_1^* z_2 + z_1^* z_\alpha - z_2 z_\alpha^*)\right],\end{aligned} \tag{B.15}$$

where

$$\gamma_M(z) = \begin{cases} z^M, & \text{for } M \geq 0,\\ (z^*)^{|M|}, & \text{for } M < 0.\end{cases} \tag{B.16}$$

The 'momentum' conservation for a particle–hole pair leads to the diagonality of the two-particle correlation function with respect to z_α. Moreover, the correlation function of the unperturbed system fulfils the following expression:

$$F^R_{0\,NM,N'M'}(\alpha, \omega) = \delta_{NN'}\delta_{MM'}F^R_{0\,NM}(\omega). \tag{B.17}$$

Consequently, the Bethe–Salpeter integral equation assumes the following form:

$$\begin{aligned}F^R_{NM,N'M'}(\alpha, \omega) = &\,\delta_{NN'}\delta_{MM'}F^R_{0\,NM}(\omega)\\ &+ \sum_{N''M''} F^R_{0\,NM}(\omega)\langle NM\alpha|W_{\mathrm{H}}|N''M''\alpha\rangle\, F^R_{N''M'',N'M'}(\alpha, \omega).\end{aligned} \tag{B.18}$$

After splitting the vertex function (B.7), it can be expressed as

$$\langle NM\alpha|W_{\mathrm{H}}^{(1)}|N'M'\alpha\rangle = q^{-2}\xi_{NM}^{(1)*}\xi_{N'M'}^{(1)} \tag{B.19}$$

and

$$\langle NM\alpha|W_{\mathrm{H}}^{(2)}|N'M'\alpha\rangle = q^{-2}\left[\xi_{NM}^{(2)*}\xi_{N'M'}^{(1)} + \xi_{NM}^{(1)*}\xi_{N'M'}^{(2)}\right], \tag{B.20}$$

where ($x = \frac{1}{2}q^2 = \frac{1}{2}|z_\alpha|^2$)

$$\xi_{NM}^{(1)} = \left[\frac{e^{-x}N!}{2\pi\bar{\rho}2^{|M|}(N+|M|)!}\right]^{1/2}\gamma_M(-z_\alpha)L_N^{|M|}(x), \tag{B.21}$$

$$\xi_{NM}^{(2)} = \left[\frac{e^{-x}N!}{2\pi\bar{\rho}2^{|M|}(N+|M|)!}\right]^{1/2}\gamma_M(-z_\alpha) \times\left[(|M|-x)L_N^{|M|}(x) + 2x\frac{dL_N^{|M|}(x)}{dx}\right]. \tag{B.22}$$

Let us define the following expression:

$$\Sigma_{ij} = \sum_{NM}\xi_{NM}^{(i)}F_{0\,NM}^{R}(\omega)\xi_{NM}^{(j)*}, \tag{B.23}$$

which corresponds to analogous sums defined for anyon superconductor (6.1):

$$\Sigma_{11} = q^2\Sigma_0, \quad \Sigma_{12} = \Sigma_{21} = q^2\Sigma_1, \quad \Sigma_{22} = q^2\Sigma_2. \tag{B.24}$$

The following expression serves to find the correlation function in the real space:

$$F^R(\mathbf{r}_1\mathbf{r}_1', \mathbf{r}_2'\mathbf{r}_2, \omega) = \sum_{NM,N'M'}\sum_{z_\alpha}\Psi_{NM\alpha}(\mathbf{r}_1,\mathbf{r}_1')F_{NM,N'M'}^{R}(\alpha,\omega)\Psi_{N'M'\alpha}^{*}(\mathbf{r}_2',\mathbf{r}_2). \tag{B.25}$$

Assuming identical space coordinates, $\mathbf{r}_1' = \mathbf{r}_1$, $\mathbf{r}_2' = \mathbf{r}_2$, the density–density correlation function for an anyon gas is obtained. One can determine

$$D_{00}^R(\mathbf{r}_1,\mathbf{r}_2,\omega) = \frac{1}{L^2}\sum_{\mathbf{q}}D_{00}^R(\mathbf{q},\omega)e^{i\mathbf{q}\cdot\mathbf{r}_{12}}, \tag{B.26}$$

where

$$D_{00}^R(\mathbf{q},\omega) = \frac{fq^2}{2\pi W}\Sigma_0. \tag{B.27}$$

In the same way, density–current and current–current correlation functions are determined and exactly the same matrix D_R^{RPA}, as in the random phase approximation is obtained. Calculating the difference between total and average currents in the Hartree approximation, the random phase approximation (RPA) is obtained (4.26).

REFERENCES

Abrikosov, A. A. (1972). *Vvedenie v teoriju normalnych metallov.* Nauka, Moskva.

Aharonov, Y. and Bohm, D. (1959). *Physical Review*, **115**, 485.

Altshuler, B. L., Ioffe, L. B., and Millis, A. J. (1994). *Physical Review*, **B50**, 14048–14064.

D'Ambrumanil, N. and Morf, R. (1989). *Physical Review*, **B40**, 6108–6119.

Arovas, D., Schrieffer, J. R., Wilczek, F., and Zee, A. (1985). *Nuclear Physics*, **B251**, 117–126.

Artin, E. (1947). *Annals of Mathematics*, **48**, 101.

Averin, D. V. and Goldman, V. J. (2002). *Solid State Communications*, **121**, 25–28.

Balachandran, A. P., Ercolessi, E., Morandi, G., and Srivastava, A. M. (1990). *Hubbard model and anyon superconductivity*, World Scientific, Singapore.

Baym, G. and Kadanoff, L. P. (1961). *Physical Review*, **124**, 287.

Benjamin, A. T., Quinn, J. J., Quinn, J. J., and Wójs, A. (2001). *Journal of Combinatorial Theory*, **A95**, 390–397.

Bennett, Ch. H. and DiVincenzo, D. P. (2000). *Nature*, **404**, 247–255.

Beran, P. and Morf, R. (1991). *Physical Review*, **B43**, 12654–12657.

Birman, J. S. (1969). *Communications on Pure and Applied Mathematics*, **22**, 41.

Birman, J. S. (1974). *Braids, links and mapping class groups.* Princeton University Press, Princeton; *Annals of Mathematics Studies*, **82**.

Bonesteel, N. E. (1999). *Physical Review Letters*, **82**, 984–987.

Bouwemeester, D., Ekert, A., and Zeilinger, A. (2000). *The physics of quantum information.* Springer-Verlag, Berlin.

Bujkiewicz, L., Jacak, L., Radosz, A., and Sitko, P. (1994). *Physica*, **B194–196**, 1469–1470.

Chen, X. M. and Quinn, J. J. (1994). *Solid State Communications*, **92**, 865–868.

Chen, Y., Wilczek, F., Witten, E., and Halperin, B. I. (1989). *International Journal of Modern Physics*, **B3**, 1001–1067.

Cowan, R. D. (1981). *The theory of atomic structure and spectra.* University of California Press, Berkeley, Los Angeles, and London.

Coxeter, H. S. and Moser, W. O. (1984). *Generators and relations for discrete groups.* Springer-Verlag, Berlin, Heidelberg, New York, and Tokyo.

Dai, Q., Levy, J. L., Fetter, A. L., Hanna, C. B., and Laughlin, R. B. (1992). *Physical Review*, **B46**, 5642–5677.

Das Sarma, S. and Pinczuk, A. (ed.). (1997). *Perspectives in quantum Hall systems.* John Wiley & Sons, New York.

Dasniéres de Veigy, A. and Ouvry, S. (1994). *Physical Review Letters*, **72**, 600–603.

Dev, G. and Jain, J. K. (1992). *Physical Review Letters*, **69**, 2843–2846.

Dieudonne, J. A. and Carrel, J. B. (1971). *Invariant theory.* Academic Press, New York .

Dirac, P. A. M. (1931). *Proceedings of the Royal Society of London*, **A133**, 60.

Dowker, J. S. (1985). *Journal of Physics*, **A18**, 3521–3530.

Du, R. R., Störmer, H. L., Tsui, D. C., Pfeiffer, L. N., and West, K. W. (1993). *Physical Review Letters*, **70**, 2944–2947.

Du, R. R., Störmer, H. L., Tsui, D. C., Yeh, A. S., Pfeiffer, L. N., and West, K. W. (1994). *Physical Review Letters*, **73**, 3274–3277.

Duda, R. (1991). *Wprowadzenie do topologii.* PWN, Warszawa.

Einarsson, T. (1990). *Physical Review Letters*, **64**, 1995–1998.

Enz, C. P. (1992). *A course on many-body theory applied to solid-state physics.* World Scientific, Singapore.

Fano, G., Ortolani, F., and Colombo, E. (1986). *Physical Review*, **B34**, 2670–2680.

Fetter, A. L. (1992). Lecture Notes, Trieste.

Fetter, A. L. and Hanna, C. B. (1992*a*). *Physical Review*, **B45**, 2335–2351.

Fetter, A. L. and Hanna, C. B. (1992*b*). *Physical Review*, **B46**, 9063–9069.

Fetter, A. L. and Walecka, J. D. (1971). *Quantum theory of many-particle systems.* McGraw-Hill, New York.

Fetter, A. L., Hanna, C. B., and Laughlin, R. B. (1989). *Physical Review*, **B39**, 9679–9681.

Fetter, A. L., Hanna, C. B., and Laughlin, R. B. (1991). *International Journal of Modern Physics*, **B5**, 2751–2790.

Fradkin, E. (1991). *Field theories of condensed matter systems.* Addison-Wesley Publishing Company, New York.

Fradkin, E., Nayak, F., Tsvelik, A., and Wilczek, F. (1998). *Nuclear Physics*, **B516**, 704–718.

Freedman, M. H., Kitaev, A., Larsen, M., and Wang, Z. (2001). *Topological quantum computation.* Preprint quant-ph/0101025.

Goldman, V. J., Su, B., and Jain, J. K. (1994). *Physical Review Letters*, **72**, 2065–2068.

Greiter, M. and Wilczek, F. (1990). *Modern Physics Letters*, **B4**, 1063.

Greiter, M., Wen, X.-G., and Wilczek, F. (1991). *Physical Review Letters*, **66**, 3205–3208.

Greiter, M., Wen, X.-G., and Wilczek, F. (1992). *Nuclear Physics*, **B374**, 567–614.

Haldane, F. D. M. (1983). *Physical Review Letters*, **51**, 605–608.

Haldane, F. D. M. (1991). *Physical Review Letters*, **67**, 937–940.

Haldane, F. D. M. and Rezayi, E. H. (1988). *Physical Review Letters*, **60**, 956–959.

Halperin, B. I. (1983). *Helvetica Physica Acta*, **56**, 75.

Halperin, B. I. and Stern, A. (1998). *Physical Review Letters*, **80**, 5457–5460.

Halperin, B. I., Lee, P. A., and Read, N. (1993). *Physical Review*, **B47**, 7312–7343.

Hamermesh, M. (1962). *Group theory and its application to physical problems.* Dover Publications, New York.

Hanna, C. B. and Fetter, A. L. (1993). *Physical Review*, **B47**, 3280–3289.

Hanna, C. B., Laughlin, R. B., and Fetter, A. L. (1989). *Physical Review*, **B40**, 8745–8758.

Hanna, C. B., Laughlin, R. B., and Fetter, A. L. (1991). *Physical Review*, **B43**, 309–319.

Hansen, V. L. (1989). *Braids and coverings.* Cambridge University Press, Cambridge.

He, S., Xie, X., and Zhang, F. (1992). *Physical Review Letters*, **68**, 3460–3463.

He, S., Simon, S. H., and Halperin, B. I. (1994). *Physical Review*, **B50**, 1823–1831.

Heinonen, O. (ed.) (1998). *Composite fermions.* World Scientific, Singapore.

Hetrick, J. E., Hosotani, Y., and Lee, B.-H. (1991). *Annals of Physics*, **209**, 151–215.

Hiller, H. (1986). *Geometry of coxeter groups.* Pitman, London and Boston.

Imbo, T. D. and Sudarshan, E. C. G. (1988). *Physical Review Letters*, **60**, 481–484.

Imbo, T. D., Imbo, C. S., and Sudarshan, E. C. G. (1990). *Physics Letters*, **B234**, 103–107.

Isakov, S. B., Lozano, G., and Ouvry, S. (1999). *Nuclear Physics*, **B552**, 677–706.

Ivanov, D. A. (2001). *Physical Review Letters*, **86**, 268–271.

Jacak, L. and Sitko, P. (1993). *Acta Physica Polonica*, **A83**, 399–416.

Jacak, D., Jacak, L., and Wieczorek, K. (1994). *Acta Physica Polonica*, **A85**, 483–496.

Jackiv, R. (1989). Lecture in *25th Anniversary Conference.* Trieste, ICTP.

Jain, J. K. (1989). *Physical Review Letters*, **63**, 199–202.

Jain, J. K. (1990). *Physical Review*, **B40**, 8079–8082.

Jain, J. K. (1994). *Science*, **266**, 1199–1202.

Jain, J. K. and Kamilla, R. K. (1997). *Physical Review*, **B55**, 4895–4898.

Jain, J. K. and Kawamura, T. (1995). *Europhysics Letters*, **29**, 321–326.

Jänich, K. (1991). *Topologia.* PWN, Warszawa.

Jones, V. F. R. (1987). *Annals of Mathematics*, **126**, 335–388.

Kadanoff, L. P. and Baym, G. (1962). *Quantum statistical mechanics.* Frontiers in Physics Series. W. A. Benjamin, New York.

Kalmeyer, V. and Laughlin, R. B. (1987). *Physical Review Letters*, **59**, 2095–2098.

Kitaev, A. V. (1997). *Russian Matematical Survey*, **52:61**, 1191–1249.

Kang, W., Störmer, H. L., Pfeiffer, L. N., Baldwin, K. W., and West, K. W. (1993). *Physical Review Letters*, **71**, 3850–3853.

Kang, M., Pinczuk, A., Dennis, B. S., Pfeiffer, L. N., and West, K. W. (2000). *Physical Review Letters*, **84**, 546–549.

Kang, M., Pinczuk, A., Dennis, B. S., Eriksson, M. A., Pfeiffer, L. N., and West, K. W. (2001). *Physical Review Letters*, **86**, 2637–2640.

Khveshchenko, D. V. (1994). *Physical Review*, **B49**, 10514–10519.

Kim, Y. B., Furusaki, A., Wen, X.-G., and Lee, P. A. (1994). *Physical Review*, **B50**, 17917–17932.

Kitazawa, Y. and Murayama, H. (1990). *Physical Review*, **B41**, 11101–11109.

von Klitzing, K., Dorda, G., and Pepper, M. (1980). *Physical Review Letters*, **45**, 494–497.

Laidlaw, M. G. and De Witt, C. M. (1971). *Physical Review*, **D3**, 1375.

Landau, L. D. and Lifszyc, E. M. (1979). *Mechanika kwantowa*. PWN, Warszawa.

Laughlin, R. B. (1981). *Physical Review*, **B23**, 5632–5633.

Laughlin, R. B. (1983*a*). *Physical Review*, **B27**, 3383–3389.

Laughlin, R. B. (1983*b*). *Physical Review Letters*, **50**, 1395–1398.

Laughlin, R. B. (1988*a*). *Science*, **242**, 525–533.

Laughlin, R. B. (1988*b*). *Physical Review Letters*, **60**, 2677–2680.

Leinaas, J. M. and Myrheim, J. (1977). *Nuovo Cimmento*, **B73**, 1–23.

Lopez, A. and Fradkin, E. (1991). *Physical Review*, **B44**, 5246–5266.

Lopez, A. and Fradkin, E. (1992). *Physical Review Letters*, **69**, 2126–2129.

Lopez, A. and Fradkin, E. (1993). *Physical Review*, **B47**, 7080–7094.

Maksym, P. A. and Chakraborty, T. (1990). *Physical Review Letters*, **65**, 108–111.

Manoharan, H. C., Shayegan, M., and Klepper, S. J. (1994). *Physical Review Letters*, **73**, 3270–3273.

Mermin, N. D. (1979). *Review of Modern Physics*, **51**, 591–648.

Milnor, J. (1968). *Singular points of complex hypersurface*. Princeton University Press, Princeton. *Annals of Mathematics Studies*, **61**.

Moore, G. and Read, N. (1991). *Nuclear Physics*, **B360**, 362–396.

Morf, R. (1998). *Physical Review Letters*, **80**, 1505–1508.

Murthy, G. and Shankar, R. (1998*a*). *Physical Review Letters*, **80**, 5458–5461.

Murthy, G. and Shankar, R. (1998*b*). In *Composite fermions* (ed. O. Heinonen). World Scientific, Singapore.

Murthy, G. and Shankar, R. (1999). *Physical Review*, **B59**, 12260–12263.

Nayak, C. and Wilczek, F. (1994). *Nuclear Physics*, **B417**, 359–373.

Nayak, C. and Wilczek, F. (1996). *Nuclear Physics*, **B479**, 529–553.

Pan, W., Du, R. R., Störmer, H. L., Tsui, D. C., Pfeiffer, L. N., and Baldwin, K. W. (1999*a*). *Physical Review Letters*, **83**, 820–823.

Pan, W., Xia, J.-S., Shvarts, V., Adams, D. E., Störmer, H. L., and Tsui, D. C. (1999*b*). *Physical Review Letters*, **83**, 3530–3533.

Pan, W., Störmer, H. L., Tsui, D. C., Pfeiffer, L. N., Baldwin, K. W., and West, K. W. (2003). *Physical Review Letters*, **90**, 016801.

Pasquier, V. and Haldane, F. D. M. (1998). *Nuclear Physics*, **B516**, 719–726.

Peshkin, M. (1981). *Physics Reports*, **80**, 375–386.

Pines, D. and Nozieres, P. (1966). *The theory of quantum liquids*. W. A. Benjamin, New York.

Polychronakos, A. P. (2000). *Physical Review Letters*, **54**, 1268–1271.

Prange, R. and Girvin, S. (ed.). (1987). *The quantum Hall effect*. Springer-Verlag, New York.

Randjbar-Daemi, S., Salam, A., and Strathdee, J. (1990). *Nuclear Physics*, **B340**, 403–447.

Read, N. (1994). *Semiconductor Science and Technology*, **9**, 1859–1864.

Read, N. (2002). *Nonabelian braid statistics versus projective permutation statistics*. Preprint hep-th/0201240.

Read, N. and Rezayi, E. H. (1999). *Physical Review*, **B59**, 8084–8092.

Rezayi, E. H. (1987). *Physical Review*, **B36**, 5454–5457.

Rezayi, E. H. and Haldane, F. D. M. (2000). *Physical Review Letters*, **84**, 4685–4688.

Rezayi, E. H. and Read, N. (1994). *Physical Review Letters*, **72**, 900–903.

Rolfsen, D. (1976). *Knots and links*. Publish or Perish, Berkeley.

Rose, M. E. (1957). *Elementary theory of angular momentum*. John Wiley & Sons and Chapman & Hall, New York and London.

Rudin, W. (1974). *Real and complex analysis*. McGraw-Hill, New York.

Schulman, L. (1968). *Physical Review*, **176**, 1558.

de Shalit, A. and Talmi, I. (1963). *Nuclear shell theory*. Academic Press, New York and London.

Shankar, R. and Murthy, G. (1997). *Physical Review Letters*, **79**, 4437–4440.

Simon, S. H. and Halperin, B. I. (1993). *Physical Review*, **B48**, 17368–17387.

Simon, S. H. and Halperin, B. I. (1994). *Physical Review*, **B50**, 1807–1822.

Sitko, P. (1992). *Physics Letters*, **A165**, 363–368.

Sitko, P. (1994*a*). *Modern Physics Letters*, **B8**, 375–380.

Sitko, P. (1994*b*). *Physics Letters*, **A188**, 179–181.

Sitko, P. (1999). *International Journal of Modern Physics*, **B13**, 2263–2274.

Sitko, P. (2001). *Physica*, **E9**, 271–274.

Sitko, P. and Jacak, L. (1994). *Physics Letters*, **A196**, 242–246.

Sitko, P. and Jacak, L. (1995). *Modern Physics Letters*, **B9**, 889–894.

Sitko, P., Yi, S. N., Yi, K. S., and Quinn, J. J. (1996). *Physical Review Letters*, **76**, 3396–3399.

Sitko, P., Quinn, J. J., and Marinescu, D. C. (1997*a*). *Physical Review*, **B56**, 3606–3608.

Sitko, P., Yi, K.-S., and Quinn, J. J. (1997*b*). *Physical Review*, **B56**, 12417–12421.

Sitko, P., Yi, K. S., and Quinn, J. J. (1998). *Physica*, **B249–251**, 32–35.

Spanier, E. (1966). *Algebraic topology*. Springer-Verlag, Berlin.

Springer, T. A. (1977). *Invariant theory*. Springer-Verlag, Berlin.

Sudarshan, E. C. G., Imbo, T. D., and Govindarajan, T. R. (1988). *Physics Letters*, **B213**, 471–476.

Tsui, D. C., Störmer, H. L., and Gossard, A. C. (1982). *Physical Review Letters*, **48**, 1559–1562.

Vladimirov, V. S. (1981). *Uravnenija matematičeskoj fiziki*. Nauka, Moskva.

Weinberg, S. (1986). *Progress of Theoretical Physics. Supplement*, **86**, 43–53.

Wen, X. G. and Zee, A. (1990). *Physical Review*, **B41**, 240–253.

Wieczorek, K. and Jacak, L. (2002). *Europhysics Letters*, **59**, 94–99.

Wilczek, F. (1982*a*). *Physical Review Letters*, **48**, 1144–1147.

Wilczek, F. (1982*b*). *Physical Review Letters*, **49**, 957–960.

Wilczek, F. (1990). *Fractional statistics and anyon superconductivity*. World Scientific, Singapore.

Wilczek, F. (1998). *Projective statistics and spinors in Hilbert space*. Preprint hep-th/9806228.

Willett, R., Eisenstein, J. P., Störmer, H. L., Tsui, D. C., Gossard, A. C., and English, J. H. (1987). *Physical Review Letters*, **59**, 1776–1779.

Willett, R. L., Ruel, R. R., West, K. W., and Pfeiffer, L. N. (1993). *Physical Review Letters*, **71**, 3846–3849.

Wójs, A. (2001). *Physical Review*, **B63**, 125312.

Wójs, A. and Quinn, J. J. (1998). *Solid State Communications*, **108**, 493–497.

Wójs, A. and Quinn, J. J. (1999). *Solid State Communications*, **110**, 45–49.

Wójs, A. and Quinn, J. J. (2000*a*). *Physical Review*, **B61**, 2846–2854.

Wójs, A. and Quinn, J. J. (2000*b*). *Physical Review*, **B62**, 4630–4637.

Wójs, A. and Quinn, J. J. (2000*c*). *Philosophical Magazine*, **B80**, 1405–1454.

Wójs, A., Hawrylak, P., and Quinn, J. J. (1998). *Physica*, **B256–258**, 490–493.

Wójs A., Hawrylak, P., and Quinn, J. J. (1999*a*). *Physical Review*, **B60**, 11661–11665.

Wójs A., Szlufarska, I., Yi, K.-S., and Quinn, J. J. (1999*b*). *Physical Review*, **B60**, 11273–11276.

Wu, Y.-S. (1994). *Physical Review Letters*, **73**, 922–925.

Wu, T. T. and Yang, C. N. (1976). *Nuclear Physics*, **B107**, 365–380.

Wu, T. T. and Yang, C. N. (1977). *Physical Review*, **D16**, 1018–1021.

Yang, S. R. E., Schliemann, J., and MacDonald, A. H. (2002). *Physical Review*, **B66**, 153302.

Yi, S. N., Chen, X. M., and Quinn, J. J. (1996). *Physical Review*, **B53**, 9599–9601.

Yoshioka, D. (1985). *Progress of Theoretical Physics. Supplement*, **81**, 97–119.

INDEX

www.ingramcontent.com/pod-product-compliance
Lightning Source LLC
LaVergne TN
LVHW011805200926
848484LV00003B/783
* 9 7 8 0 1 9 8 5 2 8 7 0 8 *